Experimental Manual in Medical Biochemistry and Molecular Biology
(Second Edition)

Chief Editors   Yu Hong   Chen Juan
Subeditors   Zhang Baifang   Du Fen   He Chunyan

图书在版编目(CIP)数据

医学生物化学与分子生物学实验指南:第二版=Experimental Manual in Medical Biochemistry and Molecular Biology(Second Edition)/喻红,陈娟主编. —武汉:武汉大学出版社,2019.11

ISBN 978-7-307-21167-4

Ⅰ.医… Ⅱ.①喻… ②陈… Ⅲ.①医用化学—生物化学—实验—医学院校—教材 ②医药学—分子生物学—实验—医学院校—教材 Ⅳ.①Q5-33 ②Q7-33

中国版本图书馆 CIP 数据核字(2019)第 203921 号

责任编辑:胡 艳　　责任校对:李孟潇　　版式设计:马 佳

出版发行:武汉大学出版社　　(430072　武昌　珞珈山)
(电子邮箱:cbs22@whu.edu.cn　网址:www.wdp.com.cn)
印刷:湖北金海印务有限公司
开本:880×1230　1/16　印张:13.25　字数:429 千字　插页:1
版次:2008 年 9 月第 1 版　　2019 年 11 月第 2 版
　　2019 年 11 月第 2 版第 1 次印刷
ISBN 978-7-307-21167-4　　定价:33.00 元

版权所有,不得翻印;凡购我社的图书,如有质量问题,请与当地图书销售部门联系调换。

Contributors

卜友泉	Bu Youquan	Chongqing Medical University, Chongqing
曹　佳	Cao Jia	Wuhan University School of Basic Medical Sciences, Wuhan
陈　娟	Chen Juan	Huazhong University of Science and Technology Tongji Medical College, Wuhan
杜　芬	Du Fen	Wuhan University School of Basic Medical Sciences, Wuhan
范成鹏	Fan Chengpeng	Wuhan University School of Basic Medical Sciences, Wuhan
过建利	Guo Jianli	Huazhong University of Science and Technology Tongji Medical College, Wuhan
何春燕	He Chunyan	Wuhan University School of Basic Medical Sciences, Wuhan
黄新祥	Huang Xinxiang	Jiangsu University, Jiangsu
季维克	Ji Weike	Huazhong University of Science and Technology Tongji Medical College, Wuhan
林　珑	Lin Long	Huazhong University of Science and Technology Tongji Medical College, Wuhan
罗洪斌	Luo Hongbin	Hubei Minzu University, Enshi
苗丽霞	Miao Lixia	Wuhan University School of Basic Medical Sciences, Wuhan
商　亮	Shang Liang	Wuhan University School of Basic Medical Sciences, Wuhan
孙续国	Sun Xuguo	Tianjin Medical University, Tianjin
田　俊	Tian Jun	Huazhong University of Science and Technology Tongji Medical College, Wuhan
武军驻	Wu Junzhu	Wuhan University School of Basic Medical Sciences, Wuhan
喻　红	Yu Hong	Wuhan University School of Basic Medical Sciences, Wuhan
张百芳	Zhang Baifang	Wuhan University School of Basic Medical Sciences, Wuhan
张瑞霖	Zhang Ruilin	Wuhan University School of Basic Medical Sciences, Wuhan
周宏博	Zhou Hongbo	Harbin Medical University, Harbin
周　洁	Zhou Jie	Huazhong University of Science and Technology Tongji Medical College, Wuhan

Preface

Progress in life science, including medical science, is mainly made by observation and experiments, especially biochemistry and molecular biology experiments. Therefore, it is essential for medical students and life science researchers to master the basic biochemical techniques and methods.

Since publication of the first edition of *Experimental Manual in Medical Biochemistry* in 2008, the pace of biotechnology has been further accelerated, the advances in bioinformatics have generated vast genomics data and subversive biotechnologies such as gene editing have emerged. While the basic biochemistry methods presented in the first edition remain as valuable tools, it is clear that graduate students and researchers need an updated reference book that incorporates classic as well as modern approaches to tackle medicinal research in functional genomics era.

We have significantly revised in the second edition in order to adapt to the recent development of new technology and research strategies in biochemistry and molecular biology. The core chapters of the previous edition were revised to highlight existing strategies and methods for nucleic acid preparation and cloning, gene transfer and expression analysis. All protocols have been evaluated, revised, and sometimes replaced with more efficient or reliable ones. New theory chapters involving the structure and function studies of gene and protein, have been introduced. New technique principles have been added, including DNA methylation technology, epigenetic analysis of chromosomal immunoprecipitation, RNAi, new generation sequencing technology, as well as technologies for protein chemical modification study, spatial structure and protein interaction research.

We thank all colleagues in the department of biochemistry and molecular biology from Wuhan University School of Basic Medical Sciences, and Huazhong University of Science and Technology Tongji Medical College, who have contributed ideas, skills and experience to this manual. We are very grateful to all the experts, especially those who helped with the preparation of the first and second editions, from other universities, who made big contribution to the compiling of various chapters. We specifically would like to thank our lovely graduate students for their contributions in drawing and proofreading. They are Guo Cui, Liu Yu, Shen Wenwen, Liu Tianli, Li Feifei, Zhao Xiaojie, Huang xiaopeng, Zhang Yu, and other students.

This manual is not a complete version. It should be continually modified and updated. We would be very grateful if the users of this manual could bring good feedbacks for the development of this manual.

Yu Hong and Chen Juan
June, 2019

Requirements

Lab safety

1. Eating, drinking, smoking or chewing gum in the laboratory are strictly prohibited.
2. Please wear lab coat and keep your work area clean. In case of spilled reagents, please wipe up as soon as possible. Be aware of objects that can burn or give electrical shocks.
3. Do not turn on an instrument until you have read instructions and consulted the instructors for the use of equipment. If any equipment malfunction is noted, report this immediately to an instructor.
4. Anyone carrying out these protocols will encounter the following hazardous materials: (1) toxic chemicals and carcinogenic or teratogenic reagents, (2) pathogens and infectious biological agents. We emphasize that users must proceed with the prudence and precaution associated with good laboratory practice. Use chemicals with high vapor pressure only in the hood. Handle and dispose of hazardous chemicals properly. Disposal containers are provided.
5. In case of an accident, notify a teacher immediately. For any chemicals splashed into the eye or mouth, or spilled on the skin, please flush immediately with large amounts of cold water using eyewash. For burns, flush with cold water and contact a teacher.

Attendance policy

Students are expected to attend every experiment, and to arrive promptly and well prepared. A student who is absent from a lab without the prior permission of the teacher or does not have documented excuse will receive 0 points on the lab report for that experiment.

Lab reports

Lab reports are due one week after completing the experiment. The reports comprise 1/2 of the practice score; the exam is 1/4, also include quiz in lab and the assessment of lab behavior.

All lab reports must be done on an individual basis. You will be given instructions about the format and the information needed for each experiment. Laboratory reports must include the following: experiment title, date, experiment performed, experimental results (primary data, calculation formulas, tables, graphs or figures, and final results), and discussion.

Content

Section I Theories and Strategies of Biochemistry and Molecular Biology Technology in Medical Research

Chapter 1 Medical Application of Biochemistry and Molecular Biology Technology ········· 3

Lin Long, Chen Juan

  1.1 Basic Objectives of Biochemistry and Molecular Biology Technology ········· 3

  1.2 Application of Biochemistry and Molecular Biology Technology in Medical Research ········· 3

Chapter 2 The Common Research Techniques of Gene ········· 8

Guo Jianli, Ji Weike, Zhang Ruilin

  2.1 Nucleic Acid Hybridization and DNA Microarray ········· 8

  2.2 DNA Sequencing ········· 11

  2.3 Polymerase Chain Reaction (PCR) ········· 13

  2.4 Recombinant DNA Technology ········· 16

  2.5 The Research Strategies and Methods for Gene Transcription Regulation ········· 22

Chapter 3 The Common Research Techniques of Protein ········· 24

Tian Jun, Zhou Jie, Fan Chengpeng, Zhang Baifang

  3.1 Quantification and Identification of Protein ········· 24

  3.2 Methods and Basic Principles of Protein Purification ········· 24

  3.3 Protein Chemical Modification and Spatial Structure Analysis ········· 26

  3.4 Protein Interaction Research Methods ········· 30

  3.5 Proteomics Research ········· 36

  3.6 Research Technology of Enzymology ········· 42

Section II Basic Biochemical Techniques

Chapter 4 Basic Operation of Biochemical Experiments ········· 47

Zhang Baifang, Wu Junzhu

  4.1 Solution Preparation ········· 47

  4.2 Pipettes Operation ········· 48

  4.3 Centrifuge Operation ········· 50

Chapter 5　Spectrophotometry ············ 52
Yu Hong, Zhou Hongbo

　5.1　Basic Concepts and Principle of Spectrophotometry ············ 52
　5.2　Measurement of Light Absorption ············ 54
　5.3　Applications of Spectrophotometry ············ 55
　Experiment 1　Quantitative Analysis of Protein ············ 58
　Experiment 2　Determination of Blood Glucose Level Regulated by Hormones ············ 66

Chapter 6　Electrophoresis ············ 70
Huang Xinxiang, He Chunyan

　6.1　Basic Principles of Electrophoresis ············ 70
　6.2　Impact Factors of Electrophoresis ············ 71
　6.3　Detection of Components after Electrophoretic Separation ············ 72
　6.4　Special Electrophoretic Techniques and Applications ············ 73
　Experiment 3　Separation of Serum Proteins by Cellulose Acetate Membrane Electrophoresis ············ 76
　Experiment 4　Separation of Serum Lipoproteins by Agarose Gel Electrophoresis ············ 79
　Experiment 5　Proteins Separation by SDS-Polyacrylamide Gel Electrophoresis ············ 82

Chapter 7　Chromatography ············ 87
Du Fen, Cao Jia

　7.1　Fundamental Principles of Chromatography ············ 87
　7.2　Commonly Used Chromatographic Techniques ············ 88
　Experiment 6　Separation of Mixed Amino Acids by Cation Exchange Chromatography ············ 94
　Experiment 7　Separation of Hemoglobin and Dinitrophenyl (DNP)-glutamate by Gel Filtration Chromatography ············ 97

Section Ⅲ　Integrative Experiments

Chapter 8　Enzyme Analysis ············ 101
Shang liang, Miao Lixia, Sun Xuguo, Zhou Jie

　8.1　The Activity of Enzyme ············ 101
　8.2　Enzyme Kinetics (Factors Affecting Reaction Velocity) ············ 102
　8.3　Enzymatic Analysis ············ 105
　Experiment 8　Assay the Activity of Alanine Aminotransferase (ALT) in Serum (Mohun's Method) ············ 108
　Experiment 9　Effect of Substrate Concentration on Enzyme Activity-Determining K_m Value for Alkaline Phosphatase ············ 111
　Experiment 10　Enzymatic Endpoint Method-Determination of Total Cholesterol, Triglycerides and Glucose Levels in Serum ············ 114

Chapter 9　Isolation, Purification and Identification of Protein ············ 121
Yu Hong, Cao Jia

　9.1　Procedure of Protein Isolation ············ 121
　9.2　The Selective Purification Steps ············ 124
　9.3　Crystallization, Concentration, Drying and Storage of Protein ············ 127

9.4　Identification and Analysis of Protein ······ 128
Experiment 11　Isolation and Identification of Serum IgG ······ 129
Experiment 12　Purification and Identification of Glutathione S-Transferase Fusion Protein ······ 135

Chapter 10　Isolation, Purification and Identification of Nucleic Acids ······ 140

Zhang Baifang, Luo Hongbin, Shang Liang

10.1　Preparation Principle and Procedure of Nucleic Acids ······ 140
10.2　Identification and Analysis of Nucleic Acids ······ 142
Experiment 13　Isolation and Identification of Eukaryotic Genomic DNA ······ 146
Experiment 14　SDS-Alkaline Lysis Preparation of Plasmid DNA ······ 154
Experiment 15　Isolation and Identification of Total RNA in Tissues/Cells ······ 157
Experiment 16　Polymerase Chain Reaction (PCR) and Reverse Transcription-PCR ······ 162

Chapter 11　Genetic Engineering ······ 167

Yu Hong, Du Fen, Bu Youquan

11.1　The Process of Genetic Engineering ······ 167
11.2　Expression of Foreign Genes and Purification of Recombinant Proteins ······ 168
Experiment 17　DNA Cloning ······ 170
Experiment 18　Expression of Exogenous Gene in *E. coli* and Purification of His-tag Proteins ······ 181

Chapter 12　Innovation Experiment Design and Application ······ 189

He Chunyan, Du Fen, Chen Juan

Experiment 19　Separation and Identification of Alkaline Phosphatase ······ 190
Experiment 20　Lactate Dehydrogenase (LDH) Analysis ······ 194
Experiment 21　Protein Markers Exploration by Two-dimensional Electrophoresis ······ 198

Reference ······ 201

Section I Theories and Strategies of Biochemistry and Molecular Biology Technology in Medical Research

Chapter 1 Medical Application of Biochemistry and Molecular Biology Technology

1.1 Basic Objectives of Biochemistry and Molecular Biology Technology

Life depends on biochemical reactions. Biochemistry is the subject of studying the chemical substances and vital processes occurring in living organisms. Biochemical research involves the structures, functions and interactions of biological macromolecules, such as proteins, nucleic acids, carbohydrates and lipids, which provide the structure of cells and perform many functions associated with life.

The double helix structure model of DNA, proposed by Watson and Crick in 1953, can be considered as the birth of molecular biology. Molecular biology is a branch of biology that concerns the molecular basis of biological processes of replication, transcription and translation of the genetic material. The central dogma of molecular biology is that genetic material which transcribe into RNA and then translate into protein, including emerging novel roles for RNA.

Biochemistry and molecular biology technologies are employed to separate and study the structure and functions of nucleic acids and proteins, to assemble recombinant DNA molecules and produce recombinant proteins, etc. In particular, medical biochemistry and molecular biology technology as an important branch is dedicated to elucidating the structure, function and expression regulation of specific molecules, and the molecular mechanisms of various physiological and pathological conditions, revealing and developing strategies and drugs for the diagnosis and prevention of different diseases.

1.2 Application of Biochemistry and Molecular Biology Technology in Medical Research

Almost all diseases have a biochemical and molecular basis. To a certain extent, all diseases are manifestations of abnormalities in genes, proteins, biochemical reactions, or metabolic processes, each of which can adversely affect one or more critical biochemical functions. New biochemical and molecular genetic tools allow investigators to query and manipulate genomic sequences as well as the entire information of cellular RNA, protein and post-translational modifications of protein at the molecular level.

1.2.1 Gene manipulation

The study of the structure and function of genes is the core of life science at the molecular level and one of the core contents of medical research. It requires many complex experimental techniques, relying on the establishment and development of technical systems. This is not a simple combination of experimental methods, but a combination of theory and technology that complements the molecular biological theory system.

In the early 1970s, techniques for nucleic acid laboratory operations appeared. These techniques, in turn, lead to the construction of DNA molecules consisting of nucleotide sequences from different sources. These innovative products, recombinant DNA molecules, have opened up exciting new pathways in molecular biology and genetics research and have given birth to a new field—recombinant DNA technology. Genetic engineering is the application of this technology in gene manipulation. It is now possible to target a specific region of almost any genome, produce an almost unlimited number of copies, and determine the sequence of its nucleotides. These technological breakthroughs in genetic engineering have a tremendous impact on the life sciences of cells and organisms.

1.2.1.1 Gene diagnosis

Gene diagnosis is the diagnosis of diseases by analyzing the mutation and expression of genes from the level of molecular structure and expression, using the current knowledge of genome and molecular genetic data.

Although much has been understood about human disease from pedigree analysis and study of affected proteins, the specific genetic defect is unknown in many cases. The development of DNA sequencing, recombinant DNA technology, high-throughput DNA microarrays, genomics analysis, and other molecular biology technology have an increasing impact on medical research and clinical medicine. Manipulation of a DNA sequence and the construction of recombinant DNA provide a means of studying how a specific segment of DNA works. Gene mapping localizes specific genes to distinct chromosomes. Gene localization thus can define a map of the human genome. This is already yielding useful information in the definition of human disease. Somatic cell hybridization and in situ hybridization are two techniques used to accomplish this.

The heritable DNA polymorphism can be associated with certain diseases in a kindred and used to search for the specific gene involved. The deletion of a critical piece of DNA, the rearrangement of DNA within a gene, or the insertion or amplification of a piece of DNA within a coding or regulatory region can all cause changes in gene expression resulting in disease.

We can use molecular approaches such as DNA probes to assist in the diagnosis of various biochemical, genetic, immunologic, microbiologic, and other medical conditions-molecular diagnostics.

1.2.1.2 Gene therapy

Gene therapy refers to the introduction of foreign normal genes into target cells to correct or compensate for diseases caused by defects and abnormal genes in order to achieve therapeutic purposes. The way to correct the defect is either to repair the defective gene in situ or to transfer the functional normal gene into a part of the cell genome to replace the defective gene. In a broad sense, gene therapy can also include measures and new technologies to treat certain diseases at the DNA level.

Since September 1990, the world's first attempt to treat adenosine deaminase deficiency-related severe combined immunodeficiency (ADA-SCID) with gene therapy has achieved gratifying results, gene therapy has made some progress in a variety of diseases, including genetic diseases, tumors, infectious diseases and so on. Because gene therapy is a new method which is different from any other therapies in the past, it is still in the early stage of development, and it is mainly aimed at diseases which have no other effective therapies in clinical routine treatment. Therefore, it will be a long time before gene therapy can be used as a routine therapy for diseases.

1.2.2 RNA study

RNA, like DNA, plays an equally important role in life activities. In most organisms, RNA is the transcription product of DNA, which participates in the replication and expression of genetic information. RNA is much smaller than DNA, but its species, size and structure are much more complex than DNA, which is closely related to its functional diversity. RNA transcript can be divided into two major categories: mRNAs encoded for proteins; the other RNAs that does not encode protein products, i.e., non-coding RNAs, including tRNA, rRNA, etc., and especially newly identified small interfering RNA (siRNA), microRNAs (miRNAs) and long non-coding RNAs (lncRNA) that have been found to play important roles in gene expression regulation in recent years, and are closely related to the occurrence and development of many diseases, such as tumors and diabetes mellitus.

1.2.2.1 MicroRNA microarray

A microRNA (abbreviated miRNA) is a class of small non-coding RNA molecule (containing about 22 nucleotides) found in plants, animals and some viruses, that functions in RNA silencing and post-transcriptional regulation of gene expression. The miRNA is partially or completely complementary to the 3′ untranslated region (3′-UTR) of the target mRNA, thereby inhibiting transcription or causing degradation of the target mRNA. Thousands of miRNAs are found in many mammalian cells, which appeared to target about 60% of genes. Many miRNAs are evolutionarily conserved, which implies their important biological functions and the relationship with disease.

High-throughput quantification of miRNAs can be carried out by microarray/chip. Be hybridized onto microarrays, slides or chips with probes to hundreds or thousands of miRNA targets, relative levels of miRNAs can be determined in different samples. By simultaneously analyzing expression level of disease-associated miRNAs and mRNAs in chip system, researchers can detect changes of miRNAs in the sample and analyze the regulatory activity of these miRNAs on target mRNAs. Some miRNA molecules have been found to be biomarkers for the early tumorigenesis, which is helpful for the early diagnosis of cancer.

1.2.2.2 The application of siRNA

The mechanism of RNA interference (RNAi) is that siRNA uses the principle of base pairing to control the stability of mRNA molecules containing sequences complementary to them. This discovery has made us a big step forward in understanding eukaryotic gene expression regulation and gene function. In addition, the strong binding of siRNA to target genes allows for their powerful ability in down-regulation of gene expression, making siRNA-based anti-disease treatment possible. Since the first discovery of RNAi in 1998, it has only been two decades, but the treatments developed based on this mechanism have entered the clinical trial stage, and many biotechnology companies specializing in the development of RNAi treatment technologies.

1.2.2.3 LncRNA microarray

Long non-coding RNAs (lncRNAs) are a class of RNAs longer than 200 nucleotides, which cannot be translated into protein. They are found to be located between coding genes or within introns. LncRNAs regulate gene expression by several different mechanisms, including gene transcription regulation, post-transcriptional regulation and epigenetic regulation.

Recognition that lncRNAs function in various aspects of cell biology has focused increasing attention on their potential to contribute towards medical research. Although many studies have implicated that aberrant expression of lncRNAs is related to a variety of human diseases, especially cancer and neurological disease (e.g., Alzheimer's disease), there is poor understanding of their role in causing disease. Recently, a number of studies examining single nucleotide polymorphisms (SNPs) associated with disease states have been mapped to lncRNAs. Many SNPs associated with certain diseases are found within non-coding regions and the complex networks of non-coding transcription within these regions make it particularly difficult to elucidate the functional effects of polymorphisms.

With lncRNA chip, researchers can detect the expression changes of lncRNAs associated with specific biological processes or diseases, rapidly in a high throughput manner. This will contribute to the subsequent lncRNAs functional studies and/or biomarker screening.

1.2.2.4 CircRNA study

Circular RNA (circRNA) is a special kind of non-coding RNA molecule. Different from traditional linear RNA (containing 5' and 3' ends), circRNA molecule has a closed ring structure and is not affected by RNA exonuclease. It is more stable and difficult to degrade. CircRNA has tissue specificity and its expression in diseases is different from that in normal cases. It can be used as a biomarker for disease diagnosis.

1.2.3 Protein study

Proteins perform most processes in cells: they catalyze metabolic reactions, use nucleotide hydrolysis to do mechanical work, and serve as the major structural elements of the cell. The great variety of protein structures and functions has stimulated the development of a multitude of techniques to study them. Protein biochemistry continues to be an essential part of modern biological research.

In order to study protein function in vitro, one must isolate a single type of protein from thousands of other proteins present in a cell. Besides some technologies for studying protein properties, structure and function, purification of protein is a basic step for characterization and study of protein. Depending on the specific properties of proteins, a series of processes and methods are used for purifying proteins. Thanks to recombinant DNA technology to produce a large amount of a given protein, protein purification has become much easier. Proteins produced in abundance by genetic engineering and protein engineering can be used for medical research, diagnostic testing, therapy (e.g., insulin, growth hormone, and tissue plasminogen activator) and vaccines preparation (e.g., hepatitis B).

The occurrence and development of diseases are related to the changes of some proteins. Protein microarray can simultaneously detect the content of all proteins in biological samples that may be related to certain diseases or environmental factors. It is also important to monitor the process of diseases or to predict and judge the effect of treatment.

If we use these proteins to construct chips to screen many candidate chemical drugs and directly screen out the chemical drugs that interact with target proteins, the development of drugs will be greatly promoted. Protein microarray is helpful to understand the interaction between drugs and their effector proteins. It can also link the action of chemical drugs with diseases, whether drugs have toxic and side effects, and determine the therapeutic effect of drugs. It can provide experimental basis for guiding clinical medication.

1.2.4 Omics

With the progress of scientific research, it is found that only one direction (genome, proteome, etc.) cannot explain all biomedical problems. Scientists propose to study the structure of human tissues and cells, the interaction between genes, proteins and other molecules from a holistic point of view, and reflect the function and metabolism of human tissues and organs through holistic analysis. Omics is the science of studying all the components of a molecule such as DNA, RNA, protein, metabolite or other molecule in a cell, tissue or the whole organism. It mainly includes genomics, proteiomics, transcriptomics and metabolomics, which have been applied to identify causal targetable pathways for disease development. These methods are appealing as these analytes are closer to phenotype development. Benefiting from the rapid development of sequencing, analytical techniques the improvement of bioinformatics platforms, omics has been widely used in medical research and had a remarkable achievement.

Genomics refers to the science of genome mapping, nucleotide sequence analysis, gene mapping and gene function analysis of all genes. The aim is to elucidate the structure, composition of genome, regulation of gene expression, the relationship between structure and function, the interaction between genes and genes, and to reveal the relationship between genes and diseases.

Transcriptomics, the transcriptome as the research object, studies all the RNA molecules in cells, and understands the temporal-spatial relationship of all transcripts produced by genes and its biological significance. In a narrow sense, transcriptomics usually refers to the study of mRNA, which is the bridge and link between genome structure and function.

Proteomics essentially refers to studying the characteristics of proteins on a large scale, including protein expression level, post-translational modification, protein-protein interaction and so on, so as to obtain a comprehensive understanding of cell metabolism, disease occurrence and other processes at the protein level.

Metabolomics is a newly developed discipline and an important part of systems biology. The concept of metabolomics comes from the metabolome. The metabolome is the complete complement of metabolites (small molecules involved in metabolism) present in a cell or an organism. Metabolomics is the in-depth study of their structures, functions, and changes in various metabolic states qualitatively and quantitatively. Unlike genomics research, genetic scores are predictive of possible changes, metabolic traits and metabolic phenotypes represent changes that are currently taking place. Therefore, metabolomics research can explore the underlying cause with a unique perspective. With the development of metabolomics detection technology, the role of metabolomics in drug discovery has become increasingly prominent. Furthermore, because metabolomics can be used for disease diagnosis and monitoring, it will undoubtedly play an important role in precision medicine. For example, given that energy metabolism abnormalities are one of the typical features of malignant tumors, the use of metabolomics methods to identify and quantify cancer metabolites may promote the development of anti-tumor drugs for energy metabolism and offer better individualized medical treatment options.

In summary, omics characterization is essential to complement genetic and biomarker studies to understand disease mechanism and therapeutics targeting.

Chapter 2 The Common Research Techniques of Gene

Gene manipulation leading to a better understanding of gene structure and function has developed modern molecular medicine. Recent advances in DNA technology have an impact on every aspect of medical diagnosis, gene therapy, forensics, and genetic profiling.

2.1 Nucleic Acid Hybridization and DNA Microarray

2.1.1 Nucleic acid hybridization

Nucleic acid hybridization technology is the fundamental tool in molecular biology. It is the process in which two single-stranded nucleic acid molecules with complementary base sequences form double-stranded nucleic acid molecules. It has been applied in various fields such as detection of gene expression, screening specific clone from cDNA or genomic library, determining the location of a gene in chromosome and diagnosis of diseases.

2.1.1.1 Principles of nucleic acid hybridization

The technique of nucleic acid hybridization is established and developed on the basis of the denaturation and renaturation of nucleic acids. Hydrogen bonds in double-stranded nucleic acids can be disrupted by some physicochemical elements, such as high temperature, and the double-stranded nucleic acids are separated into two component strands. If the renaturation is allowed to occur under proper conditions, for example, annealing at the lower temperature, different single-stranded DNA molecules, or DNA and RNA molecules, or RNA molecules mixed together in a solution, will pair with each other to form a local or whole molecule of double-stranded structure as long as the single-stranded molecules are complementary no matter what kind of sources they come from. This is called hybridization which indicates that each strand of DNA came from a different source.

The high specificity of base pairing interaction between complementary strands of DNA can be used to locate specific nucleotide sequences in a sample. The DNA from one source can be immobilized by attachment to a solid surface. Pairing DNA sequence from another source will hybridize to the immobilized DNA. This is the basis of DNA probe technique. A probe is a specific DNA or RNA sequence that is labelled with a marker which allows for identification and quantification. A probe labelled with a detectable tracer is the prerequisite for determining a target DNA sequence or gene in a sample or genomic DNA by nucleic acid hybridization. The target nucleic acids are usually denatured and mixed with the labelled probe in the hybridization system. The probe will bind to the segment of nucleic acid with complementary sequence under proper conditions, thus the existence or the expression of the specific gene can be determined by the detection of the tracer labelling probe.

2.1.1.2 Preparation and labelling of nucleic acid probe

The working probe should be single-stranded molecules. The probes used in nucleic acid hybridization include genomic DNA fragment, cDNA fragment, oligonucleotide (15~50nt) and RNA. Genomic DNA probes can be prepared from a cloned DNA fragment in plasmid. cDNA probes can be prepared from a cloned cDNA in plasmid, or amplified directly from mRNA by RT-PCR. Oligonucleotide probes are short single-stranded DNA fragments designed with a specific sequence complementary to the given region of the target DNA. They are usually synthesized in vitro. RNA probes are usually transcribed in vitro from a cloned cDNA in a proper vector. The size of genomic DNA probes, cDNA probes and RNA probes may be 0.1kb to 1kb.

DNA or RNA probes are usually labelled with a detectable tracer, including radioactive labelling and non-radioactive labelling. Radioactive labelling has very high sensitivity and specificity, but it is harmful to our bodies and contaminates the environment. Non-radioactive labelling is safety for us, and it is also a very stable labelling which can be kept for over one year, while its disadvantage is low sensitivity and specificity.

The labels in common use include radioactive (^{32}P and ^{35}S) and nonradioactive (digoxigenin, biotin, fluorescein) substances which are used to label dNTP. There are two types of labelling methods. One type is called end labelling which puts the labels at the ends and the other is known as uniform labelling which puts the labels internally. For end labelling, the purified oligonucleotide is usually labelled in vitro by using a suitable enzyme to add the labelled nucleotide to the end of the oligonucleotide.

Genomic DNA probes and cDNA probes are usually labelled in the process of DNA synthesis in vitro. It is socalled uniform labelling. In the reaction of DNA synthesis with a DNA probe as template, if a labelled-NTP, which can be incorporated into newly-synthesized DNA chain, is added as a substrate, the labelled DNA probe will be formed. There are different, sensitive detecting methods for each of the labels used in nucleic acid hybridization. After hybridization, the location and the quantity of the hybrid molecules can be determined.

Digoxigenin and biotin can normally bind to the specific nucleotide of the probe. When labelled probe binds to the DNA sequence to be detected, the digoxigenin or biotin can be recognized by its specific antibodies conjugated with fluoresce or enzyme such as alkaline phosphatase, AP or horseradish peroxidase, HRP. Then the DNA bands can be detected through substrate application and AP or HRP catalytic reaction (Fig. 2-1).

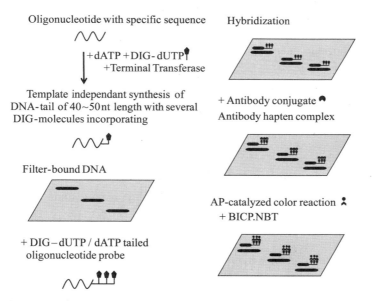

Fig. 2-1 Nonradioactive labelled probes produced by PCR DIG-labelling reaction

2.1.1.3 Different types of hybridization

1. Southern blot

Southern blot is an assay for sample DNA by DNA-DNA hybridization which detects target DNA fragments that have been size-fractionated by agarose gel electrophoresis. The basic procedure of Southern blot is that the target

DNA fragments is digested with one or more restriction endonucleases, size-fractionated by agarose gel electrophoresis, denatured with alkali, such as NaOH, then transferred (blotted) to a nitrocellulose (NC) or nylon membrane for probe hybridization and finally detected (Fig. 2-2).

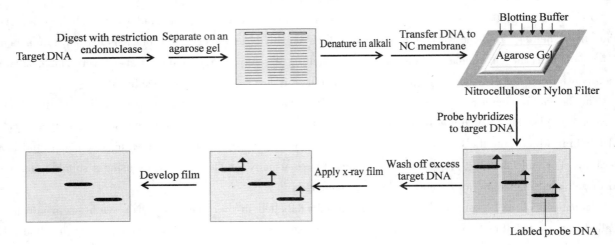

Fig. 2-2 The basic procedures of Southern blot

The individual denatured DNA fragments immobilized on the membrane are allowed to interact with labelled probe and form a target-probe DNA heteroduplex, therefore, the positions of the immobilized target DNA on the membrane are faithful records of the size separation achieved by agarose gel electrophoresis, they can be related back to the original gel to estimate their size.

Southern blot technique is widely applied in research field since it was invented by a scientist called Southern EM. It could be applied for analysis of gene expression, screening of recombinant plasmids or phages, analysis of gene mutation, restriction fragment length polymorphism (RFLP), and identification of the existence of a foreign DNA such as DNA from pathogenic microorganism. It could also be used to detect deletion of gene by restrictions mapping.

2. Northern blot

The main principle of Northern blot is similar to that of Southern blot, but the target nucleic acid is RNA instead of DNA. There are some differences between Northern blot and Southern blot: ① RNA must be denatured before agarose gel electrophoresis, while DNA is not denatured before and during electrophoresis; ② After electrophoresis, RNA in gel can be transferred to NC membrane directly, while DNA in gel must be denatured with alkali and neutralized before transfer; ③ RNA-DNA hybridization is not as strong as DNA-DNA hybridization. The hybridization solution for RNA-DNA hybridization contains more substances to accelerate the RNA-DNA binding.

3. Dot blot and slot blot

The principle of dot blot and slot blot hybridization is the same as that of Southern blot and Northern blot, while the procedure is much simpler. Dot blot and slot blot hybridization are suitable approaches if the purpose is only to determine mRNA expression quantitatively.

The general procedure of dot or slot blot involves taking an aqueous solution of target DNA or RNA, which is simply spotted on to a NC or nylon membrane as dot or slot without gel electrophoresis. In both methods the target DNA sequences are denatured, either by exposing to heat or alkali condition. The denatured target DNA or mRNA sequences immobilized on the membrane are exposed to a solution containing single-stranded labelled probe sequences. Then they are allowed to form probe-target heteroduplex and detect by exposed to an X-ray film.

2.1.2 DNA microarrays

DNA microarrays (gene chips) contain a series of DNA sequences attached to a glass slide in a precise order, the slide is divided into a grid of microscopic spots, or "wells". Each spot contains thousands of copies of a particular oligonucleotide of 20 or more bases. A computer controls the addition of these oligonucleotide sequences in a predetermined pattern. Each oligonucleotide can hybridize with only one DNA or RNA sequence and thus is a unique identifier of a gene. Thousands of different oligonucleotides can be placed on a single microarray.

Microarrays provide large arrays of sequences for hybridization experiments. It can be used to examine patterns of gene expression in different tissues and under different conditions, also used to identify individual organisms with particular mutations.

2.2 DNA Sequencing

DNA sequencing is a key technology for analyzing gene structure and understanding gene function. In 1977, dideoxy chain termination method was first established by F. Sanger, and in the same year, the chemical cleavage method was also developed using chemical degradation technique for DNA sequencing.

2.2.1 Dideoxy sequencing (Sanger's method)

The principle of dideoxy chain termination method described by F. Sanger for DNA sequencing is that DNA synthesis is terminated by the use of dideoxynucleoside triphosphate (ddNTP). It consists of a catalyzed enzymatic reaction that polymerizes the DNA fragments complementary to the template DNA of interest (unknown DNA). Briefly, a ^{32}P-labelled primer (short oligonucleotide with a sequence complementary to the template DNA) is annealed to a specific known region on the template DNA, which provided a starting point for DNA synthesis. In the presence of DNA polymerases, polymerization of deoxynucleoside triphosphates (dNTP) for DNA elongation is continued until the enzyme incorporates a modified ddNTP (called a terminator) into the growing chain. When it is added, chain elongation stops because there is no 3'-OH for the next nucleotide to be attached to. By setting the concentration of the ddNTP to be much lower than that of its normal dNTP, chain termination will occur randomly at one of many positions containing any ddNTP.

This method was performed in four different tubes, each containing the appropriate amount of one of the four terminators. All the generated fragments had the same 5'-end, whereas the residue at the 3'-end was determined by the ddNTP used in the reaction. After all four reactions were completed, the mixture of different-sized DNA fragments was separated by electrophoresis on a denaturing polyacrylamide gel electrophoresis (PAGE) in four parallel lanes. The pattern of bands showed the distribution of the termination in the synthesized strand of DNA. Following electrophoresis, an autoradiography film is developed, and the unknown sequence could be read off by reading the bands from the bottom to the top, a direction that gives the 5'→3' sequence of the newly synthesized DNA which is complementary to the provided DNA template (Fig. 2-3).

2.2.2 Maxam-Gilbert chemical sequencing

The chemical method of DNA sequencing was developed by A. Maxam and W. Gilbert. The method is based on the base specific chemical cleavage taking place in DNA molecule. While still using polyacrylamide gels to resolve DNA fragments, the Maxam-Gilbert technique differed significantly in its approach from Sanger method. Instead of relying on DNA polymerase to generate fragments, radiolabelled DNA is treated with chemicals which break the chain at specific bases. In this method, four sets of oligodeoxynucleotides are generated by subjecting a purified 3'-end or 5'-end labelled oligodeoxynucleotide to a base specific chemical reagent that randomly cleaves

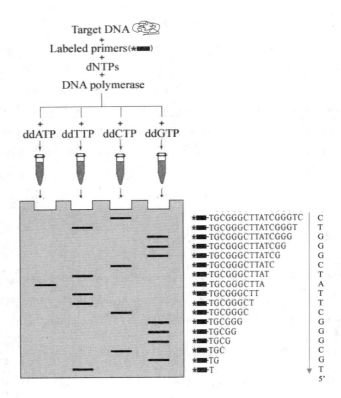

Fig. 2-3 Principle of dideoxy sequencing.

DNA at one or two specific nucleotides. Only end-labelled fragments can be observed following autoradiography of the sequencing gel.

The method works for both single-stranded and double-stranded DNA and involves base-specific cleavages which occur in two steps. In first step, DNA molecules are labelled with radioactive phosphor (^{32}P) at only one end (5′-end) with the help of a special enzyme, polynucleotidekinase and separated into four different reactions. Each chemical reaction cleaves the DNA preferably at only one point using a mild nucleotide-specific treatment. The results are a mixture of DNA fragments of different lengths (between one and several hundred nucleotides), with one end representing one of the bases and the other end showing the ^{32}P phosphate. The DNA fragments are then applied on parallel lanes on a sequencing gel and separated. The result is a specific pattern, which shows the nucleotide sequence after applying autoradiography.

The two sequencing methods mentioned above were widely adopted, and thus might be considered the real birth of "first-generation" DNA sequencing.

2.2.3 Automated DNA sequencing

Sequencing technology based on Sanger sequencing is still the gold standard and many large and small laboratories which have access to this technology. Large-scale DNA sequencing efforts are dependent on improving efficiency by partial automation of the techniques involved. One major improvement has been the development of automated procedures for fluorescent DNA sequencing. These procedures generally use 4 kinds of ddNTPs labelled different chemical fluorophores. During electrophoresis, a monitor detects and records the fluorescence signal as the DNA passes through a fixed point in the gel. The use of different fluorophores in the four base-specific reactions means that all four reactions can be loaded into a single lane with capillary electrophoresis (CE). The output is in the form of intensity profiles for each of different fluorophores, and the information is simultaneously stored electronically. Recent advances in technology mean that the accuracy of DNA sequencing using automated methods is acceptably high.

2.2.4 Second and third generation of DNA sequencing

With the advent of genome era, traditional sequencing methods can no longer meet the requirements of large-scale genome sequencing, so the second and third generation sequencing technologies have emerged one after another.

New techniques markedly differed from existing methods in that it did not infer nucleotide identity using radio- or fluorescently—labelled dNTPs or oligonucleotides before visualizing with electrophoresis. Instead, researchers utilized a recently discovered luminescent method for measuring, for example, pyrophosphate synthesis. In the main, these new technologies move away from Sanger sequencing (electrophoresis) to modular unitary base addition onto multiple templates arrayed onto a solid phase, using microfluidics and new approaches to sequence addition and ultra-sensitive signal reading technologies to capture the signal from the densely arrayed reactions ($10^4 \sim 10^6$ features per cm^2), and thus bring about enormous increases in productivity and volume of quality data.

Compare to the first-generation of DNA sequencing, the second-generation of DNA sequencing has some characteristics and advantages: ①The second-generation DNA sequencing can realize high throughput and large scale sequencing due to the use of the array and parallel sequencing, so it can save time and manpower; ②The process of second-generation DNA sequencing does not need electrophoresis to separate DNA fragment, so the machine can be made smaller; ③ This technology can also save the cost of the sequencing due to the obvious reduction consumption of samples and chemicals. The main disadvantage of second-generation DNA sequencing is that the reliable reads is short between 30bp to 450bp.

The third-generation sequencing technology, such as HeliScope single-molecule sequencing technology, is characterized by single-molecule sequencing. It does not need to treat the samples for PCR amplification, which not only avoids the possibility of errors caused by amplification, but also reduces the use of reagents and greatly reduces the cost of sequencing. It can detect the number and sequence structure of DNA/RNA molecule more intuitively, thus making the detection of nucleic acid molecules more accurate.

2.3 Polymerase Chain Reaction (PCR)

PCR is a revolutionary method which was originally developed in 1983 by the American biochemist Kary Mullis. The principles underlying DNA replication in cells have been used to develop this important laboratory technique and then it allows researchers to make a huge number of copies of a specific DNA sequence *in vitro*. PCR has revolutionized molecular genetics by permitting rapid cloning and analysis of DNA. There have been numerous applications in molecular cloning, diagnosis of genetic diseases, forensic science, and archaeology, etc.

2.3.1 The principles of PCR

PCR is used to amplify a segment of DNA that lies between regions of known sequences. PCR is an enzymatic reaction *in vitro* in the presence of DNA polymerase, template DNA, primers and 4 dNTPs. The two oligonucleotides are used as a pair of primers for synthetic reactions which catalyzed by a DNA polymerase. The template DNA is firstly denatured by heating in the presence of two primers and dNTPs. The reaction mixture is then cooled to a temperature that allows the primers to anneal to their target sequences, after which the annealed primers are extended with DNA polymerase. The cycle of denaturation, annealing and DNA synthesis is then repeated many times. Because the products of one round amplification serve as templates for the next, each successive cycle essentially doubles the amount of the desired DNA product.

2.3.2 The basic steps of PCR

The PCR amplification is a cyclic process in which a sequence of steps is repeated for 25 or 40 cycles. The cycles are done on an automated cycler, a device which rapidly heats and cools the test tubes containing the reaction mixture. The three steps include denaturation, annealing and extension and take place at a different temperature:

(1) Denaturation: To denature DNA, it must be heated to more than 90℃. At 95℃, the double-stranded DNA melts and opens into two pieces of single-stranded DNA (ssDNA).

(2) Annealing: At medium temperatures, around 54℃, the primers pair up (anneal) with the ssDNA "template". On the small length of double-stranded DNA (dsDNA) which joins primer and template, the DNA polymerase attaches and starts to copy the template.

(3) Extension: At 72℃, DNA polymerase works at the highest activity, and dNTPs complementary to the template are coupled to the primer, making a dsDNA molecule. With one cycle, a single segment of dsDNA template is amplified into two separate pieces of dsDNA. These two pieces are then available for amplification in the next cycle. This three-step process is repeated more and more copies are generated and the number of copies of the template is increased exponentially (Fig. 2-4).

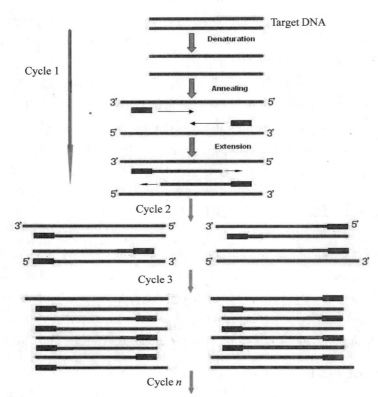

This three-step process is repeated for a number of cycles (usually 20~30 cycles), resulting in the production of many copies (10^6~10^9) of the template DNA sequence.

Fig. 2-4 The principle of PCR

2.3.3 PCR reaction system

In nature, most organisms copy their DNA in the same way. The PCR mimics this process, and only it is done in a test tube. The PCR reaction mixture contains:

2.3.3.1 Template DNA

Any source of DNA that provides one or more target molecules can in principle be used as the template of PCR. The DNA templates may be DNA fragments, genomic DNA, recombinant DNA, recombinant plasmid or λ phages. And the form of template DNA can be single or double-stranded, closed circular DNA or linear DNA. Whatever the source of template DNA is, the target segment of DNA can only be amplified under the precondition that some sequence information is known so that primers can be designed. High concentration of template DNA can leads to non-specific amplification. In order to guarantee specific amplification, 1μg DNA genomic DNA and about 10ng of plasmid DNA are commonly suggested to be used for one PCR reaction.

2.3.3.2 A pair of primers

Primer is artificially synthesized DNA and complementary in sequence to a specific region of the template DNA. A pair of primers, including forward and reverse primers, are usually synthesized according to the amplified DNA sequence. A variety of computer programs are available to aid in PCR design. The general design principle of primer is to raise the efficiency and specificity of PCR, including: ① the length of 15~30 base is suitable; ② G + C content of primer should be 45%~55%; ③ 3′ end of primer must be matched with the DNA template; ④ It should not contain complementary sequences between or in primers, avoiding primer dimers; ⑤ It has no strict limitations for 5′ terminal of primer, which can be in a free state, not matching with the DNA template. Therefore, the primers can be designed with the RE sites or other short sequences (such as ATG initiation code or a mismatched base mutation, etc.) at the 5′ end.

At present, computer-assisted retrieval analysis has been widely used in the design of primers, and the gene sequences can be found from Genebank or EMBL data-bank.

2.3.3.3 DNA precursors

The four deoxynucleoside triphosphates (dNTPs) include dATP, dCTP, dGTP and dTTP. They can be obtained either as freeze-dried or neutralized aqueous solutions. The suitable concentrations of dNTPs between 50~200mmol/L each result in the optimal balance among yield, specificity and fidelity.

2.3.3.4 DNA polymerase

DNA polymerase synthesizes new strands of DNA complementary to the target sequence. The first and most commonly used enzyme is *Taq* DNA polymerase, isolated from a bacterium called *Thermus aquaticus*, which lives in the very hot water of geothermal vents. One drawback is its lack of 3′~5′ exonuclease (proof reading) activity, which can lead to the misincorporation of nucleotides. Another polymerase is *Pfu* DNA polymerase (from *Pyrococcus furiosus*), which is used widely because of its higher fidelity when copying DNA.

2.3.3.5 Reaction buffer

To maintain a near neutral pH in PCR reaction system, the recommend buffer for PCR is 10mmol/L Tris buffer, with a pH range between 8.5 and 9.0 at 25℃. This value is optimal for the activity of *Taq* polymerase.

2.3.3.6 Magnesium ions

The magnesium concentration may affect all of the following: primer annealing, strand dissociation, product specificity, formation of primer-dimer artifacts, and enzyme activity and fidelity. Because the dNTPs bind magnesium ions, the reaction mixture must contain an excess of Mg^{2+} (0.5~2.5mmol/L greater than the concentration of dNTPs).

2.3.4 Derivatives techniques of PCR

Since its birth, PCR technology has been continuously developed and improved, and now a variety of PCR technologies have been derived. Here are some commonly used techniques.

2.3.4.1 Nested PCR

Nested PCR is developed to increase the sensitivity of PCR by using a second pair of primers to amplify the product from the primary PCR. For this purpose, two pairs of primers are designed, with the second pair of primers placed internal to the first pair of primers. The product amplified with first pair of primers is used as template for the amplification with second pair of primers.

2.3.4.2 In situ PCR (ISP)

In situ PCR is conducted in formalin fixed and paraffin-embedded tissue slice or cell smear, then in situ hybridization is conducted by specific probe, thus determining whether DNA or mRNA exists in the investigated tissue or cell. ISP is the combination of PCR and in situ hybridization. It has many advantages over conventional PCR and in situ hybridization, as it overcomes the limitation that PCR can not be able to determine the location of specific DNA or mRNA, and can detect DNA or mRNA expressed at a low level due to its great accuracy and sensitivity.

2.3.4.3 Reverse transcription-PCR (RT-PCR)

RT-PCR is a highly sensitive technique for detection of gene expression and quantitation of mRNA. The technique consists of two parts: ① the synthesis of cDNA from RNA by reverse transcription (RT); ② The amplification of a specific cDNA by the PCR.

RT-PCR uses RNA as starting material for nucleic acid amplification in vitro. Reverse transcriptase catalyzes cDNA synthesis using RNA as the template, and cDNA is not subject to RNase degradation, making it more stable than RNA. In RT-PCR, the starting RNA is subsequently degraded, dsDNA is produced, and PCR amplification proceeds in the usual manner.

2.3.4.4 Real-time PCR

Real-time PCR also known as quantitative PCR (qPCR), was developed to analyze the amount of PCR product quantitatively. It monitors the amplification of a targeted DNA molecule during the PCR, i.e. in real-time, and not at its end, as in conventional PCR. The products of real-time PCR are analyzed in the process of the reaction, and the results are generated more quickly with greater accuracy and reliability. It is increasingly being adopted for RNA quantification and genetic analysis. This exciting technology has enabled the shift of molecular diagnostics toward a high-throughput, automated technology with lower turnaround times.

2.4 Recombinant DNA Technology

DNA recombination is the process that at least two DNA fragments from different sources (cell, bacteria or even species) join together and form recombinant DNA. In 1973, Herbert Boyer and Stanley Cohen successfully created the recombinant DNA technology. They isolated two different plasmids from Escherichia coli (*E. coli*), cut with REs, and rejoined a recombinant plasmid with DNA ligase. The ligation products were transformed into *E. coli* and the transformant containing the target genes were grown on the medium, so the result was amplification of a particular gene or DNA segment. The methods used to accomplish these and related tasks are collectively referred to recombinant DNA technology, also called gene cloning, DNA cloning or genetic engineering. In 1976, Genentech Inc. founded by Boyer et al., produced the first genetic engineering product—insulin.

The recombinant DNA technology makes it possible to isolate and identify a gene, and induce the gene to express functional protein in host cells. The genetic endowment of organisms can now be precisely changed in designed ways. The development of recombinant DNA techniques has revolutionized biology and is having an increasing impact on clinical medicine. It offers a rational approach to understanding the molecular basis of a number of diseases, for example, new insights are emerging into the regulation of gene expression in cancer. At present, clinically therapeutic proteins almost are produced by recombinant DNA technology. This powerful technique is also applied in diagnosis or gene therapy. In addition, this technology paves the way for the modern fields of genomics and proteomics. Many genetically modified organisms have been developed and approved for use in agriculture, medicine, and other industries.

2.4.1 How to clone a gene

One goal of recombinant DNA technology is to clone a target gene, that is, to produce many identical copies of a particular gene or DNA segment. This process is also called DNA cloning. The basic procedures of DNA cloning include: ① obtaining the target gene and selection of a suitable vector; ② digestion and ligation of the target gene and vector; ③ transformation of the recombinant DNA into host cells; ④ screening the positive host cells harboring recombinant DNA; ⑤ expression and identification of the recombinant DNA (Fig. 2-5).

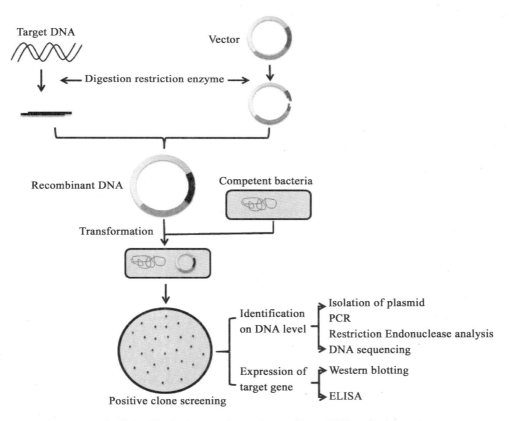

Fig. 2-5 The basic procedures of recombinant DNA technology

2.4.1.1 Preparation of the target gene

A major goal of DNA cloning is to study the functions of a target gene or DNA sequences and the proteins they encode. The interested DNA fragments can be obtained from several sources. They include random DNA fragments of chromosome that are maintained as gene libraries, cDNA obtained by reverse transcription from mRNA, products of PCR, and artificially synthesized or mutated DNA.

1. Libraries provide collections of DNA fragments

A genomic library is a collection of DNA fragments that comprise the genome of an organism. The cDNA library is a collection of cDNAs from a particular tissue at a particular time in the life cycle of an organism. In fact, a genomic library or cDNA library of an organism is constructed by the process of DNA cloning. Each clone of host cells in genomic library or cDNA library carries a special DNA fragment. cDNA is also a good starting point for cloning eukaryotic genes, because the clones contain only the coding sequences of the gene. A single target DNA can be isolated from the library by a variety of screening protocols.

2. The target DNA amplified by PCR

If the sequence of DNA fragment is known, the target DNA fragment can be prepared by PCR. RT-PCR can also be used to create and amplify a specific cDNA sequence without the need to construct a library. In this case, RNA is isolated from an organism or tissue, cDNA will be made from mRNA using reverse transcriptase, then PCR is used to amplify a specific sequence from cDNA.

3. DNA synthesized by organic chemistry

Artificial DNA can be synthesized by DNA synthesizer using organic chemistry to link nucleotides together in a specific sequence. The target DNA is usually short-length sequence.

2.4.1.2 Selection and preparation of a suitable vector

Molecular cloning involves the amplification of a given DNA molecule by replication in a host cell. However, the DNA fragment to be amplified does not have an origin of replication, it must be inserted into a carrier which can be replicated in bacterial cells and passed to subsequent generations of the bacteria. Such a carrier, termed "vector", is usually a DNA molecule which has a suitable origin of replication such as plasmid DNA, bacteriophage and other virus DNA. If a vector is used for amplification of a DNA fragment, it is called "cloning vector". If a vector is used for the expression of a given gene in the host cell, it is called "expression vector". The selection of vector is dependent on the purpose of the construction of recombinant DNA.

1. Cloning vectors

The ideal cloning vectors are well-characterized molecules of DNA and display the following properties: ① they are not integrated into the host genome so that they are easily isolated and purified; ② it is easy for them to enter into host cells because of their small size; ③ they contain replication origins that allow them to be replicated independently in host cell; ④ they contain one or more selectable markers such as antibiotic resistance gene that allow the easy identification of the positive host cells; ⑤ they have multiple single restriction sites called multiple cloning site (MCS) which foreign DNA can be inserted in.

2. Expression vectors

Expression vectors require not only transcription but also translation of the cloned gene, thus require more components than simpler cloning vectors. In addition to having common vector features such as *ori.*, MCS and selectable marker(s), expression vectors should contain a strong promoter and translational control sequences (Fig. 2-6 and Table 2.1). The cloned gene should contain or be engineered to contain an open reading frame (ORF), which starts with a translation initiator ATG and stops with a stop codon. The promoter is positioned immediately upstream of the MCS where the foreign gene is inserted. It drives the transcription of the cloned gene, and often constitutes the major determinant for the degree of expression level obtained. The translational control sequences such as ribosomal-binding site are required for the efficient translation of the transcribed mRNA. The combination of the promoter, cloned gene and translational control sequences is often referred to as an "expression cassette". Moreover, in some vectors, tag sequences are introduced into the expression vector to facilitate the detection and purification of the recombinant proteins (Table 2.1).

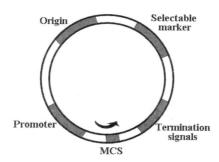

Fig. 2-6 The general structure of expression vectors

Table 2.1   Features of expression vectors

Features	Prokaryotic expression vector	Eukaryotic expression vector
Selectable marker	Ampicillin resistance Tetracycline resistance Kanamycin resistance	Neomycin resistance Puromycin resistance Blasticidin resistance
Promoter	Trp-lac T7 Phage T5	CMV β-actin EF-1α
Translational control sequences	Shine-Dalgarno (SD) sequence	Kozak sequence Polyadenylation signal

2.4.1.3 Ligation of DNA molecules

After obtaining target gene and selection of suitable vector, the following step needed to be done is to cut them and join the two different fragments together. The discovery of two types of enzymes permitted the common technique of DNA cloning. One type of enzymes, called restriction enzymes from bacteria, can cut DNA from any organism at specific sequences of a few nucleotides, generating a reproducible set of fragments. The other type of enzymes, called DNA ligases, can catalyze the joining of DNA restriction fragments, generating recombinant DNA.

1. Restriction endonucleases (REs)

There are three general types of REs. Type I and III cleave DNA away from recognition sites. Only Type II REs can cleave DNA at specific recognition sites, and can be used to prepare DNA fragments containing specific genes or portions of genes in gene cloning techniques. For example, the RE *Eco*R I, isolated by Herbert Boyer in 1969 from *E. coli*, cleaves the DNA between G and A in the base sequence 5'GAATTC3'. The main advantage of REs is their ability to cut DNA strands reproducibly in the same places. This property is the basis of many techniques used to analyze genes and their expressions. Moreover, many REs make staggered cuts in the two DNA strands, leaving sticky ends, which can form base pair together briefly. Some enzymes cut in the middle of its recognition sequence, so it produces blunt ends with no overhangs.

2. DNA Ligase

DNA ligase is a particular type of ligase that can link together DNA strands that have double-stranded breaks (a break in both complementary strands of DNA). It catalyzes the formation of a phosphodiester bond between the

5′ phosphate termini of one strand of duplex DNA and the 3′ hydroxyl termini of the other. This enzyme will join blunt end and cohesive end termini as well as repair single stranded nicks in duplex DNA, RNA or DNA/RNA hybrids and is extensively used to covalently link fragments of DNA together. Most commonly, it is used with restriction enzymes to insert DNA fragments, often genes, into plasmids vector, which is a fundamental technique in recombinant DNA work.

After digestion with REs, any two pieces of DNA containing the same sequences within their sticky ends can anneal together and be covalently ligated together in the presence of DNA ligase. DNA fragments digested by the same REs generally will link together through base-pairing so that it is very easy for DNA ligase to fill the nicks. Any two blunt-ended fragments of DNA can be ligated together irrespective of the sequences at the ends of the duplexes.

The simplest strategy for cloning is thus to cut the donor and vector DNA with the same restriction enzyme and join them together with DNA ligase. However, this allows the vector to reclose without an insert. There are two procedures which prevent vector self-ligation: ① the liner vector fragment can be treated with alkaline phosphatase to remove 5′ phosphate groups of the vector so that the liner vector cannot reclose; ② the vector and donor can be prepared by digestion with same pair of restriction enzymes so that both vector and donor will have two incompatible ends, but the compatible ends are existent between vector and donor DNA. A further advantage of the second strategy is that the orientation of the insert can be predicted (directional cloning).

3. Other Enzymes for Gene Cloning

Alkaline phosphatase is used to remove 5′-phosphate group necessary for ligation to avoid the self-ligation of vectors. Klenow fragment is the large fragment of DNA polymerase I, it is used for the conversion of both 5′ and 3′ protruding termini to blunt ends. RNase H is an endoribonuclease that specifically hydrolyzes the phosphodiester bonds of RNA which is hybridized to DNA. RNase H and reverse transcriptase are both enzymes for making cDNA libraries.

2.4.1.4 Introduction of recombinant DNA into host cells

After recombinant DNA molecules are formed, they need to be introduced into host cells, which amplify the DNA molecules in the course of many generations of cell division. A number of techniques are available, such as chemical sensitization of cells, electroporation and biolistics. Plasmid DNA can be introduced into bacterial cells such as *E. coli* by transformation. If the host cells are mammalian cells, the process is known as transfection, usually using calcium-phosphate-based transfection, liposome mediated transfection, microinjection, or infection through the use of viral particles.

Many bacteria, including *E. coli*, do not seem to have evolved or retained ways of taking up functional DNA. The treatment with Ca^{2+} induces a transient state of "competence" in recipient bacteria (*E. coli*), during which the bacteria possess the capacity of taking up foreign DNA. For example, *E. coli* treated with $CaCl_2$ make the cell wall more permeable to DNA, these cells are called competent cells.

2.4.1.5 Screening and Identification of recombinant DNA

Neither DNA manipulation nor gene transfer procedures are 100% efficient, so it is desirable to identify the recombinant population. This process is termed screening or selection. There are direct and indirect systems employed depending on the vector (Fig. 2-7).

The gene or genotype can be detected directly based on some genetic markers of vector and target gene. The usual simple strategy is to check the selective markers, such as resistance to an antibiotic. Cloning vector often contains genes that confer resistance to different antibiotics (ter^r, amp^r or kan^r). Only cells transformed by the plasmid can survive and grow in the presence of that antibiotic. Additionally, the cloning vectors may contain color

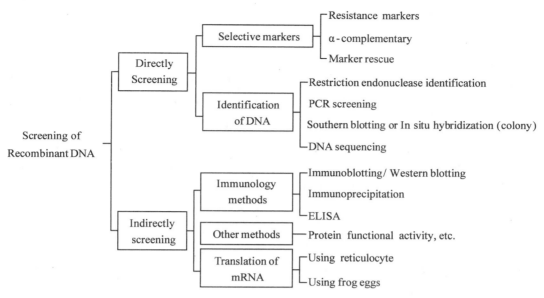

Fig. 2-7 DNA recombinant screening methods

selection markers which provide blue/white screening (α-factor complementation) on X-gal medium. Nevertheless, these selection steps do not absolutely guarantee that the DNA insert is correctly present in the cells obtained. Further investigation of the resulting colonies is required to confirm whether the cloning is successful. This may be accomplished by means of PCR, restriction fragment analysis and/or DNA sequencing, etc.

If a gene has been cloned into an expression vector, the best selective step is to check the corresponding protein expressed by the target gene in bacteria, yeast or mammalian cells. The selection of the system depends on biochemical and biological properties of the target protein, the requirements for functional activity and the desired yield.

2.4.2 Expression of the cloned gene

In many cases, gene cloning is employed to express the desired protein for subsequent applications such as functional analysis of novel proteins and therapeutic uses. An expression system mainly consists of an expression vector, its cloned gene, and the host cells which provide the necessary conditions to allow the cloned gene express. Expression vectors have transcriptional promoters immediately adjacent to the site of insertion.

There are two expression systems: prokaryotic expression system and eukaryotic expression system. Generally, the choice of expression system is determined by the particular properties of the protein to be expressed and the purpose for which it is intended.

The typical prokaryotic expression system is *E. coli* expression system. In most cases, this system is simple, fast, economical and suitable for large scale production of target proteins. However, *E. coli* expression system also has some disadvantages especially when eukaryotic genes are expressed in *E. coli*. The disadvantages of *E. coli* expression system including: ① lack of post-transcription processing. Only cDNA of a cloned eukaryotic gene can be expressed using this system; ② lack of appropriate post-translational processing. The eukaryotic proteins cannot be properly folded or glycosylated in *E. coli*; ③ production of insoluble inclusion body of expressed protein. It is often difficult to get large amount of soluble protein products through this system. If protein products are soluble, *E. coli* cells frequently recognize the foreign protein and destroy them by proteolytic degradation.

Specialized expression vectors have been designed to improve the efficiency of *E. coli* expression system. Most vectors are inducible so that premature overproduction of a foreign product that could poison the bacterial host cells can be avoided. Expression vectors often contain a short sequence coding for a tag (e.g., six histidines tag, His-tag) just upstream of the multiple cloning site to produce fusion proteins, a fusion product of the tag and the target protein. Then the fusion proteins can be readily purified by affinity chromatography (e.g., Ni-NTA affinity chromatography for His-tag fusion proteins).

Eukaryotic systems have some advantages over their prokaryotic counterparts for producing eukaryotic proteins. At first, eukaryotic proteins made in eukaryotic cells tend to be folded properly, so they are soluble, rather than aggregated into insoluble inclusion bodies. Secondly and more importantly, eukaryotic proteins made in eukaryotic cells are usually properly modified (phosphorylated, glycosylated, etc.) in a eukaryotic manner, and they almost always accumulate in the correct cellular compartment. However, compared with prokaryotic expression system, eukaryotic expression system is technically difficult, time-consuming, expensive, and relatively slow because eukaryotic host cells are more difficult to manipulate, which have higher demand on media and grow slower than prokaryotic host cells.

2.5 The Research Strategies and Methods for Gene Transcription Regulation

The basic mechanism of gene transcription regulation is on the interaction between DNA and proteins. Transcriptional regulatory proteins, also known as trans-acting factors, bind to the corresponding DNA cis-acting elements and are the basis for chromatin remodeling and gene transcriptional regulation. Therefore, the identification of proteins capable of binding to DNA regulatory elements is the main idea for elucidating the regulation mechanism of gene expression. Techniques that can be used for this purpose include gel electrophoresis mobility analysis, chromatin immunoprecipitation, oligodeoxynucleotide precipitation, and so on.

2.5.1 Gel migration or electrophoretic mobility experiment

Electrophoretic mobility shift assay (EMSA), also known as gel mobility shift assay, is based on the mobility shifts of macromolecules by electrophoresis to show protein-DNA binding. The migration of DNA fragments combined with protein molecules in the gel will lag behind. The interaction between proteins and corresponding DNA sequences can be used for qualitative and quantitative analysis and has become a classical method for transcription factor research. The disadvantage is that EMSA can only indicate that DNA sequence contains protein binding sites, but it cannot determine the specific sequence.

2.5.2 Chromatin immunoprecipitation

Chromatin immunoprecipitation (ChIP) is another common technique for studying the interaction of proteins and DNA. Unlike EMSA, ChIP technology can truly reflect the binding of proteins to genomic DNA *in vivo*. Living cells were treated with formaldehyde to cross-link proteins with DNA. Next, Chromosomes were separated and randomly broken into small chromatin fragments (200~2000bp). Nucleoprotein-DNA complexes were precipitated by specific antibodies to target proteins, and then DNA fragments were amplified and identified by PCR. Its application range has been developed from the study of the interaction between the target protein and the known DNA target sequence to the study of the interaction between the target protein and the unknown DNA sequences in the entire genome. The technology also extends from the study of a single target protein and DNA interactions to study protein complexes that bind to DNA sequences. This technique also allows for the identification of site-specific histone modifications of the promoter region.

2.5.3 Promoter activity analysis

The transcriptional activity of the promoter refers to the ability of the gene to initiate transcription, which is affected by many factors such as the structure of promoters, the presence of transcription factors and the presence of other regulatory proteins. The transcriptional activities of the promoter *in vivo*, including the transcriptional activity changes under various environmental signals, are mainly reflected by the quantitative analysis of transcript RNA. Currently, the most commonly used technology for quantitative analysis of RNA transcripts is real-time quantitative PCR technology (RT-PCR). The *in vivo* environment is complex, and it is difficult to study the transcriptional initiation activity of specific promoters, especially for studies of specific cis-acting elements. Therefore, *in vitro* analysis systems are mainly used to study transcriptional regulation of specific genes. Promoter activity analysis requires the cloning of specific genes and their promoters and examine the effects of various factors on their transcription in host cells. The transcriptional activity of the promoter still requires detection of the RNA transcript. The analysis and detection methods of most RNA transcript are complex, rendering transcriptional regulation research difficult to perform. Some new technologies have solved this problem to some extent, which has greatly promoted transcriptional regulation research, such as reporter gene technology, Nuclear run-on transcription assay, rapid amplification of cDNA end (RACE), and so on.

2.5.4 DNA methylation analysis

DNA methylation refers to the selective addition of methyl group to cytosine in DNA under the catalysis of methyltransferase to form 5'-methylcytosine, which is commonly found in 5'-CG-3' sequence in genome. Most vertebrate genomic DNA has a small amount of methylated cytosine, which is mainly concentrated in the 5-terminal non-coding region of the gene and exists in clusters. Methylation sites can be inherited as a result of DNA replication since methylases can methylate newly synthesized unmethylated sites after DNA replication. DNA methylation plays a key role in the development of higher organisms, the maintenance of gene expression patterns, and the stability of the genome. A large number of studies have shown that DNA methylation is closely related to gene inactivation. In general, methylation of the promoter region of a gene results in the closure of gene expression, or silencing. Changes in methylation status are an important factor in tumors, including a decrease in the overall methylation level of the genome and an abnormal increase in the local methylation level of the CpG island, resulting in genomic instability, such as chromosome instability, mobile genetic factors, expression of proto-oncogenes, and inactivation of tumor suppressor genes. Therefore, CpG methylation has become one of the research hotspots in current molecular biology. In recent years, DNA methylation detection and analysis technology has been greatly developed and improved. At present, a wide range of applications include methylation-specific PCR (MSP), sulfite-modified genome sequencing (BSP), methylation-sensitive restriction endonuclease, and bisulfite-binding restriction endonuclease method.

Chapter 3   The Common Research Techniques of Protein

Proteins are the most abundant class of biological macromolecules as they represent over 50% of the dry weight of cells. They are involved in virtually all chemical reactions and processes occurring in life forms as well as in the mechanical support and filamentous architecture within and between cells. Irrespective of their functional or structural role, they all are formed from the same 20 amino acids, which are joined end-to-end through covalent linkages termed peptide bonds. The size, ionization properties and hydrophobicity/ hydrophilicity of the constituent amino acids influence the solubility, stability and structural organization of proteins that exist essentially in either aqueous or membrane environments. The amino acid sequences (primary structures) determine the higher structural levels (secondary and tertiary) of proteins and specify their biological properties and physicochemical properties: ① protein has zwitterionic properties; ② proteins have colloidal properties; ③ many factors can cause protein denaturation; ④ the protein has a characteristic absorption peak at 280nm; ⑤ The protein color reaction can be used for the determination of protein in solution. These properties of protein can be used for the separation, purification and identification of protein.

Protein biochemistry is an essential part of modern biological research. The increase in genome sequencing facilitates gene cloning, protein production, and the study of protein properties, structure and function. Tremendous advances in heterologous expression of recombinant proteins have greatly increased our ability to produce proteins of interest. The separation and purification of protein provide the basis for further research on the structure and function of proteins.

3.1   Quantification and Identification of Protein

Accurate protein quantitation is essential to all experiments related to proteins in a multitude of research topics. Different methods have been developed to quantitate proteins in a given assay for total protein content and for a single protein. Total protein quantitation methods comprise traditional methods such as the measurement of UV absorbance at 280nm, Bicinchoninic acid (BCA) and Bradford assays, as well as alternative methods like Lowry or novel assays developed by commercial suppliers (see Chapter 5). Typically commercial suppliers provide a well-designed, convenient kit for each type of the assays.

Individual protein identification methods include enzyme-linked immunosorbent assay (ELISA), Western blot analysis (see Chapter 9), and mass spectrometry. Validation of protein expression typically involves an assessment of expression and solubility of the target protein and qualitative verification of the expected protein size. The expression level of most target proteins can be detected by denatured gel electrophoresis. Proteins expressed at low levels or that may be marginally soluble may require a more sensitive detection method such as Western blotting.

3.2   Methods and Basic Principles of Protein Purification

Protein purification is vital for the characterization of the structure, function, post-translational modifications and interactions of the proteins of interest. Prior to beginning the purification of a protein, one must generally consider what the protein is to be used for, In particular, how about is properties and sensitivities of the target protein, how much of the amount of protein is needed, what should be its state of purity and activity. Using such criteria, the appropriate methods should be chosen and a procedure be planned.

Protein purification is a series of processes intended to isolate one or a few proteins from a complex mixture, usually cells, tissues or whole organisms. The preparation of protein is generally divided into the following stages: ① pretreatment and cell separation; ② cell lysis and cell organelle separation; ③ extraction, separation and purification of proteins; ④ quantification, concentration, drying and preservation (see Chapter 9). In the purification process, special attention must be given to the preservation of the protein integrity and prevent the loss of biological activity of the proteins due to acid, alkali, high temperature and violent mechanical action.

Based on the protein's physicochemical properties, such as its size, shape, isoelectric point (pI), solubility and hydrophobicity, and some of its chemistry, one kind of protein can be isolated and purified from a complex mixture using a group of reasonable separation methods (Table 3.1). Usually more than one purification step is necessary to achieve the desired purity.

Table 3.1 **Methods for protein separation**

Physicochemical Properties	Methods for Separation
Solubility	Precipitation by salting out (e.g., ammonium sulfate); Acetone; pI; PEG
Size and shape	Gel-electrophoresis; Gel filtration; Centrifugation; Dialysis and ultrafiltration
Density	Ultracentrifugation
Charge	Gel-electrophoresis; Ion exchange chromatography
Isoelectric point (pI)	Isoelectric focusing electrophoresis; chromatofocusing
Hydrophobicity	Hydrophobic interaction chromatography (HIC); Reverse phase-HPLC
Biospecific interaction	Affinity chromatography

3.2.1 Separation by precipitation

Precipitation with salts such as ammonium sulfate at high ionic strengths or with organic solvents (e.g. ethanol) at pI is a very common initial step in protein purification. This operation also serves to concentrate and fractionate the target product from various contaminants. It has the advantages of economy, simplicity and high capacity.

3.2.2 Dialysis and ultrafiltration

Dialysis is the process of separating molecules in solution by the difference in their rates of diffusion through a semi-permeable membrane. Ultrafiltration is a technique related to dialysis, using high-pressure by gas pressure or centrifugation, to force water and other small molecules through the selectively permeable membrane, the proteins are retained and concentrated in the container. Dialysis and ultrafiltration are typically used to separate proteins from buffer components for buffer exchange, desalting, or concentration.

3.2.3 Ultracentrifugation

Ultracentrifugation is valuable for separating biomolecules and investigating such parameters as mass and density by calculating their sedimentation coefficient (S), and learning about the shape of a molecule. The most common density gradient ultracentrifugation is used to separate proteins with different S, especially useful for separation of lipoproteins.

3.2.4 Separation by electrophoresis

Proteins are ampholytes having a pH-dependent net charge in a solution. Using electrophoresis in different ways, it is possible to separate proteins, detect protein contamination, estimate molecular weight (MW) and pI and obtain information on the physiological activity of the protein (see chapter 6). Usually the SDS-Polyacrylamide Gel Electrophoresis (SDS-PAGE) methods are used in protein separation. SDS-PAGE is rapid, sensitive, and capable of a high degree of resolution, normally be used for checking the efficacy of protein purification scheme. Isoelectric focusing (IEF) is a high resolution electrophoretic technique that can be used for preparative work as well as for analytical studies. Proteins are separated according to their charges (isoelectric points, pI).

3.2.5 Purification by chromatography

Most purification schemes involve high resolution chromatographic techniques such as ion exchange chromatography, gel filtration, affinity chromatography, et al. (see Chapter 7)

Ion-exchange chromatography separates proteins based on the reversible ionic interaction between a charged protein and an oppositely charged ion exchanger which is either polyanions or polycations. Size exclusion chromatography separates molecules on the basis of their size and shape. Affinity chromatography separates protein based on a reversible interaction between a protein and a specific ligand attached to a chromatographic matrix (such as that between antigen and antibody, enzyme and substrate, or receptor and ligand). Hydrophobic interaction chromatography separates proteins with differences in hydrophobicity.

3.2.6 Concentration, drying and preservation of protein

Protein is diluted due to column purification during the preparation process. For the purpose of preservation and identification, concentration is often required. Usually it is necessary to concentrate a sample following gel filtration, and can improve the sensitivity of subsequent analysis. In addition, proteins are generally more stable in relatively concentrated solutions. Some methods can be employed as a means of concentration, such as precipitation, concentrative dialysis and ultrafiltration. The optimal concentration method depends on the volume of the sample, the type of protein, the type of buffer employed and the aim of preparation.

Purified proteins often need to be stored for an extended period of time while retaining their original structural integrity and/or activity. Protein can be preserved in dry powder and liquid form. The principles of protein preservation are low temperature, split charging, concentrating, quick freezing/quick melting, additives. The choice of preservation method of protein depends entirely on the stability of protein and when to reuse the protein after preservation.

Drying protein is a dehydration process typically used for long-term storage or more convenient for transportation. Normally, the methods used for protein drying have vacuum drying and freeze drying (also known as lyophilization).

3.3 Protein Chemical Modification and Spatial Structure Analysis

3.3.1 Protein chemical modification

Protein chemical modification is covalent additions introduced to amino acid side chains or termini of proteins, either enzymatically or chemically. It is also called post translational modification (PTM) as a protein undergoes a chemical modification after its translation, including removal and/or derivatization of specific amino acid residues, proteolytic cleavage, loss of signal sequences, and formation of disulfide cross-links. Other common post-translational modifications are phosphorylation, glycosylation, acetylation, hydroxylation, myristoylation, isoprenylation, and many other reactions, such as spontaneous oxidation or deamidation. All amino acid side chains, except alanine, glycine, isoleucine, leucine, and valine, can be modified.

Most commonly used detection method of known modifications is through immuno/Western blotting. This method depends on distinctive antibody to recognize the known modification. However, the methods are not available for all PTMs and cannot distinguish enzyme-specific modification events. The bottom-up proteomics (i.e., shotgun proteomics), which is based on mass spectrometry (MS) analysis of proteolytic peptides, has become the most powerful technique for in-depth characterization of protein PTMs. The general shotgun proteomics workflow for analysis of protein PTMs mainly includes the steps of protein digestion, specific PTM peptide enrichment, pre-fractionation and LC separation, and MS detection.

However, the identification and characterization of PTMs are still limited by the poor knowledge of the underlying enzymatic reactions and their final effects on protein stability and dynamics. In silico methods, often based on the current knowledge of PTMs, are a promising strategy to perform preliminary analysis and prediction that can guide further *in vivo* and *in vitro* experiments, leading to expand our understanding of the role of PTMs in cellular processes. Several computational approaches have been developed to study PTMs (e.g., phosphorylation, sumoylation or palmitoylation) showing the importance of these techniques in predicting modified sites that can be further investigated with experimental approaches. The existing computational approaches used to study the most common PTMs classified based on the covalent attachment of small chemical groups, lipids or small proteins to the main peptide chain. Most of these tools are presented as online webservers, providing a user-friendly interface for PTM site identification.

3.3.2 Determination of the three-dimensional (3D) structure of a protein

The three-dimensional structure of a biological macromolecule constitutes not only the basis for investigating its function but can also serve as a template to determine how and where small molecule ligands bind to it. Such an approach has proven to be extremely useful for the identification and characterization of putative lead compounds that affect the function of a macromolecular target, and at some point, it may lead to the development of new drugs. The direct way to determine the 3D structure of a protein include X-ray crystallography, nuclear magnetic resonance (NMR) methodologies, and Cryo-electron microscopy (Cryo-EM). The mg amounts of pure proteins are needed for these techniques. It is also essential that the purity level is as high as possible (>95%). This includes both the chemical purity (if other molecules still present in the mixture) and the conformational purity (does the molecule to be crystallized occur in different conformational states).

3.3.2.1 X-Ray crystallography

In this method, the crystals of a pure protein are exposed to an X-ray beam. X-ray is diffracted by the lattice of atoms present in a protein crystal. The diffraction pattern of the X-ray depends on the number of electrons in the atoms and the organization of atoms in a protein molecule. The diffracted X-rays detected as reflections on a detector could be used to produce an electron density map. This map is used to generate the picture of atoms in a protein molecule to provide the 3D structure of that protein. X-ray crystallography has been used to elucidate the 3D structure of macromolecules, including proteins and nucleic acids.

One major difficulty of this methodology lies in getting the protein in pure form and then in generating the crystals of protein molecules. Proteins are difficult to be obtained in crystal forms. It is more of an art than a science to produce the crystals of proteins. Proteins that are present in low abundance, and particularly the hydrophobic proteins or proteins with hydrophobic segments such as the membrane proteins, are difficult to crystallize. These proteins are not amenable to the analysis by X-ray crystallography. Two methodologically distinct methods can be applied in protein crystallization, the batch method and the diffusion based method. In the batch method, the protein and the precipitant solution are mixed and placed in a sealed container so that no further transport process takes place. In the diffusion-based methods, the crystallization experiment starts in a stable solution. Depending on the experimental design, there may be some differences in the path of crystallization. Nowadays, the hanging-drop set-up is the most popular and widespread approach. In this technique, the hanging drop contains the protein and some precipitant in a defined volume ratio while the reservoir solution only contains the precipitant. The solutions are separated by a vapor phase and the whole system is sequestered by sealing it with silicon oil or grease. The system is initially not in thermodynamic equilibrium because of the differences between the precipitant concentration in the reservoir and that in the drop. Eventually an equilibrium is achieved by the diffusion of water and other volatile components (if present) from the drop to the reservoir. Other diffusion-based techniques are the sitting-drop technique, which also makes use of diffusion across the vapor phase, dialysis, which utilizes diffusion across a membrane, diffusion across an oil barrier, diffusion across a phase boundary when the protein solution and the precipitant solution are placed next to each other without mixing them or diffusion through gels.

In addition to the difficulty of crystallization, X-ray crystallography generates a tremendous amount of data that must be analyzed to generate the 3D picture of the protein molecule. The 3D structure of proteins can also be determined by neutron diffraction. A protein crystal is exposed to the neutron beam, and the position of atoms in a protein is determined by the scattered neutron. Unlike X-ray diffraction, the neutron is scattered by the nucleus of the atom and not by electrons; therefore, this method yields a different kind of picture at the atomic level than X-ray crystallography. Only a small number of proteins have been analyzed by neutron scattering. One major problem of this method is that only a handful of neutron scattering devices are available in the world. Thus, fewer than a dozen proteins have been analyzed at the atomic level by this method.

3.3.2.2 Nuclear Magnetic Resonance (NMR) Spectroscopy

The principle of NMR spectroscopy is based on the magnetic properties of the nuclei of certain atoms. Atoms that contain odd number(s) of protons or (H1) or certain stable isotopes, such as H2, C13, N15, P31, and F19, with an odd number of protons or neutrons behave like a magnet. When exposed to radio waves of a certain frequency in an external magnetic field, these atoms absorb the radio waves of the frequency equal to the frequency of their spin; this phenomenon is called "resonance". This perturbation in the state of the atom is also called "chemical shift" and is used to determine the chemical nature of the atom. After absorption of radio waves, the atoms enter an excited stage but later emit the radiation equal to the amount radiation absorbed. The amount of emitted radiation and the time taken to emit the radiation can be measured and used to understand the nature of the atom. NMR has a low resolution with a low signal-to-noise (S/N) ratio, even though the NMR signal provided some information about the nucleus of an atom. This difficulty of NMR was overcome by the application of Fourier transformation by Robert Ernst, and soon, 2D and multidimensional NMR became available. Later, Kurt Wuthrich developed NMR suitable for the analysis of the 3D structure of a protein molecule.

Hydrogen (H1) is the most abundant atom in a protein. NMR signals for H1 are obtained to decipher the 3D structure of a protein. The NMR signals for H1 vary in different positions; for example, the NMR signal for H in CH3 is different than in CH or OH. Also, the NMR signal for H attached to C is different from the H attached to N. In addition, two adjacent H atoms bound by covalent bonds provide different NMR signals than when they exit as noncovalently bound or as single H atoms occurring alone or remote from each other. Now, it is feasible to determine the different kinds of NMR signals for H1 because of the availability of multidimensional NMR spectroscopy. Essentially, three different kinds of NMR signals are generated. These include Nuclear Overhauser effect spectroscopy (NOESY), correlation spectroscopy (COSY), and total correlation spectroscopy (TOCSY).

NMR signals from atoms that are close in space but not linked by covalent bonds are called "the Nuclear Overhauser Effect". This technique is useful in deciphering the positions of atoms in the structure of a protein where many distant atoms are brought together in space because of the folding of polypeptide segments, assuming the secondary and tertiary structure of the protein. COSY identifies the positions of atoms joined by chemical bonds. TOCSY identifies the atoms that are part of the networks not necessarily connected by any chemical bond. All these different measurements establish the distances among different atoms and neutrons behave like a magnet and possess a spin. Thus, atoms of hydrogen are used to decipher the 3D structure of the protein. In addition to obtaining the information about the position of hydrogen atoms, it is necessary to obtain more information about the carbon and nitrogen atom. To do this, proteins are labelled with C13 and/or N15 using the cloned gene to express labelled proteins in bacteria. The labelled proteins are then examined by NMR to locate the position of the C and N atoms in the protein.

NMR is usually useful for determining the structure of protein that is available in a solution, although recent advances have led to the development of solid NMR spectroscopy. NMR can determine the structure of proteins under physiological conditions. This methodology is also useful in determining the structure of proteins under denaturing conditions. NMR is helpful in deciphering the 3D structure of small proteins of molecular weights up to 35,000 daltons; however, through recent advances in this methodology, it is possible to determine the structure of a large protein. NMR has been adapted for imaging purposes in medicine, which include revealing images of internal organs for medical diagnosis, revealing a disease, or discovering physiological alterations in human. This adapted form of NMR is called "magnetic resonance imaging" (MRI).

3.3.2.3 Cryo-electron microscopy

Cryo-electron microscopy (Cryo-EM) is a powerful tool to determine the three-dimensional structure of biological macromolecular and cells. The core of it is transmission electron microscopy (TEM), by using direct electron detectors and advanced image-processing algorithms, the resolution can reach near-atomic level. The basic processes include sample preparation, TEM imaging, image processing and structure analysis. In TEM imaging, electrons generated by electron guns are accelerated to sub-light speed in high voltage electric field and move inside the high vacuum microscope. According to the deflection principle of high speed electrons in magnetic field, a series of electromagnetic lenses in TEM converge electrons, focus and image the electrons witch interact with the sample in the process of penetrating. Finally, the sample images magnified hundreds of times are formed on the recording medium, two-dimensional projection of three-dimensional objects can be obtained from multiple angles, and the fine three-dimensional structure of biological samples can be reconstructed by computer.

According to the properties and characteristics of different biological samples, different methods of microscopic imaging and three-dimensional reconstruction can be adopted in the practice of structure analysis, including single particle analysis (SPA) and subtomogram averaging (STA). Cryo-electron microscopy (TEM) single-particle technique can be used to study the structure and conformation changes of biological macromolecules in solution, without crystallization, relatively small amount samples (0.1 ~ 2mg/mL). It is suitable for studying > 80 ku proteins, viruses and other biological macromolecules and their complexes. The larger molecular weight and higher symmetry samples are easier to reconstruct the high-resolution structure.

A unique specimen preparation method was adopted to preserve the biological specimens at near native condition within a thin amorphous ice film. The sample particles are suspended in glassy ice to maintain the near-natural high-resolution structure of biological molecules and reduce the irradiation damage by electron beam. Liquid samples are applied to a holey carbon film covering a special EM grid which is conductive materials (copper, gold, etc.). The hydrated specimen is added onto the perforated grid, which is pre-treated by glow discharging, covered with a filter paper to remove excess solution under tightly controlled environmental conditions, then specimen plunged rapidly into liquid ethane, which is generally precooled with liquid nitrogen. So under liquid nitrogen temperature or below, the specimens can be directly observed on a low dose transmission electron microscope operating, and do not require 3D crystals. Macromolecules conformations or biomolecular complex can be observed directly in their native environment. The method has become even more widely applicable with present-day single particle analysis and electron tomography.

Biomolecule has weak electron scattering ability and low imaging contrast. In order to reduce radiation damage, biological samples are imaged with low electron dose (usually about $20e/Å^2$), resulting in lower signal-to-noise ratio and contrast. Generally, a large number of particles need to be collected. The higher is the target resolution, the more particles are needed. The data collected depends on sample symmetry, homogeneity, imaging quality and target resolution. At present, it usually takes about 50,000 to 150,000 particles to achieve near-atomic resolution by direct electronic detection camera imaging for asymmetric molecules.

The electron microscopic image of each particle needs the translation degree of freedom in the plane, orientations in the plane and orientations away from the plane. The translation degree of freedom can be eliminated by alignment. The rotational degrees of each particle leaving the plane provide three-dimensional information of the particle. The purpose of single particle image processing is to calculate and establish these parameters of each particle by using the reconstruction software of electron microscope. The commonly used reconstruction software includes EMAN2, SPIDER, XMIPP, IMAGIC, etc. The principle of 3D reconstruction from 2D projections is based on the central section theorem.

It is difficult to reconstruct the high resolution structure of some samples, especially small molecular and poor homogeneity, even so cryo-EM can provide valuable structural information for unknown samples. Biomolecule structure is flexible and changeable, it may have different conformations or assembly modes under different functional states. It is very important to understand the molecular dynamics related to function by observing and screening these structural changes by cryo-EM single particle technology. It can not only analyze the molecular framework but also provide molecular dynamic information.

3.3.2.4 Bioinformatics approach

Many mathematical methods are developed to build a homology model based on the 3D structural model of a closely related protein (if there is one). There are various servers that will perform the process largely, or completely, automatically. The best known is SWISS-MODEL. This accepts a protein sequence and will return a 3D model if it is able to build one. More advanced users can submit multiple sequence alignments and manually refine the final model. It is important to remember that any homology-built model will, at best, be imperfect and at worst totally misleading—particularly if one or more of the structural models that act as a template for the model contain errors. So a key part of SWISS-MODEL is the various validation checks applied to each model to provide the user with an idea of its likely quality.

3.4 Protein Interaction Research Methods

Typically proteins hardly act as isolated species while performing their functions in vivo. It has been revealed that over 80% of proteins do not operate alone but in complexes. Protein-protein interactions (PPIs) are the physical contacts of high specificity established between two or more protein molecules as a result of biochemical events steered by electrostatic forces including the hydrophobic effect. PPIs handle a wide range of biological processes, including cell-to-cell interactions and metabolic and developmental control. Aberrant PPIs are the basis of multiple aggregation-related diseases, such as Creutzfeldt-Jakob, Alzheimer's diseases, and may lead to cancer.

The study of PPIs is important to infer the protein function within the cell. The functionality of unidentified proteins can be predicted on the evidence of their interaction with a protein, whose function is already revealed. The detailed study of PPIs has expedited the modeling of functional pathways to exemplify the molecular mechanisms of cellular processes. At present, protein-protein interaction (PPI) is one of the key topics for the development and progress of modern system's biology.

Noncovalent contacts between the residue side chains are the basis for protein folding, protein assembly, and PPI. These contacts induce a variety of interactions and associations among the proteins. Based on their contrasting structural and functional characteristics, PPIs can be classified in several ways. On the basis of their interaction surface, they may be homo-or heterooligomeric; as judged by their stability, they may be obligate or nonobligate; as measured by their persistence, they may be transient or permanent. A given PPI may be a combination of these three specific pairs. The transient interactions would form signaling pathways while permanent interactions will form a stable protein complex.

PPIs have been studied from different perspectives: biochemistry, quantum chemistry, molecular dynamics, signal transduction, among others. All this information enables the creation of large protein interaction networks-similar to metabolic or genetic/epigenetic networks-that empower the current knowledge on biochemical cascades and molecular etiology of disease, as well as the discovery of putative protein targets of therapeutic interest.

Protein-protein interaction detection methods are categorically classified into three types, namely, *in vitro*, *in vivo*, and in silico methods. The *in vitro* methods include tandem affinity purification, affinity chromatography, co-immunoprecipitation, protein arrays, protein fragment complementation, phage display, X-ray crystallography, and NMR spectroscopy. The *in vivo* methods in PPI detection are yeast two-hybrid (Y2H, Y3H) and synthetic lethality. In silico techniques are performed on a computer (or) via computer simulation. The in silico methods in PPI detection are sequence-based approaches, structure-based approaches, chromosome proximity, gene fusion, in silico 2 hybrid, mirror tree, phylogenetic tree, and gene expression-based approaches. The diagrammatic classification was given in Table 3.2. This article only briefly introduces co-immunoprecipitation, affinity chromatography, yeast two-hybrid, intracellular co-localization analysis and protein interaction prediction method.

Table 3.2 **Summary of PPI detection methods**

In vitro	In vivo	In silico
TAP-MS	Yeast 2 hybrid(Y2H)(H)	Ortholog-based sequence approach
Affinity chromatography	Synthetic lethality	Domain-pairs-based sequence approach
Coimmunoprecipitation	Co-localization fluorescent staining	Strture-based approaches
Protein microarrays(H)	Fluorescence resonance energy transfer(FRET)	Gene neighborhood
Protein-fragment complementation	Surface Plasmon resonanc(SPR)	Gene fusion
Phage display(H)		In silico 2 hybrid(12H)
X-ray crystallography		mirrotree
NMR spectroscopy		correlated mutation

3.4.1 Co-immunoprecipitation(co-IP)

Co-immunoprecipitation (co-IP) is a popular technique for protein interaction discovery. Co-IP is an effective method to determine the physiological interaction of two proteins in intact cells. Co-IP is conducted in essentially the same manner as immunoprecipitation (IP) of a single protein, except that the target protein precipitated by the antibody, also called the "bait", is used to co-precipitate a binding partner/protein complex, or "prey", from a lysate.

The basic principle of co-IP is to harvest and lyse cells while maintaining protein interaction, add a specific antibody against a known protein to the cell lysate, and then add protein A/G-agarose beads that can bind to the antibody to precipitate and harvest the antigen-antibody complex. If there is a target protein bound to the known protein in the cell, it can be precipitated together with the above antigen-antibody complex to form a "target protein-known protein-anti-known protein antibody-protein A" complex. After SDS-PAGE, the complex can be separated and the target protein can be identified by western blot or mass spectrometry.

This method is often used to determine whether the two proteins bind *in vivo*. Antibodies against the two proteins are often used to carry out co-IP respectively to confirm each other. In combination with mass spectrometry technology, it can also be used to identify a new unknown binding protein for a specific protein. The main steps of co-IP are basically the same as immunoprecipitation (Fig. 3-1).

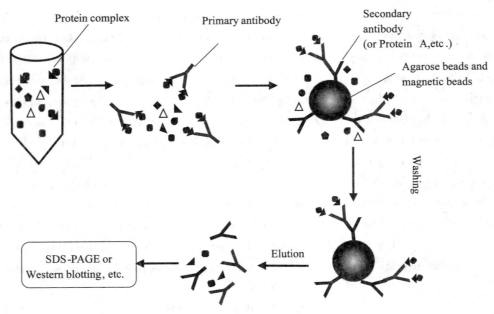

Fig. 3-1 General schematic of co-IP

3.4.2 Affinity chromatography (pull-down assay)

A pull-down assay is a small-scale affinity purification technique similar to immunoprecipitation, except that the antibody is replaced by some other affinity system. In this case (Fig. 3-2), the affinity system consists of a glutathione S-transferase (GST)-, polyHis- or streptavidin-tagged protein or binding domain that is captured by glutathione-, metal chelate (cobalt or nickel)- or biotin-coated agarose beads, respectively. The immobilized fusion-tagged protein acts as the "bait" to capture a putative binding partner (i.e., the "prey"). In a typical pull-down assay, the immobilized bait protein is incubated with a cell lysate, and after the prescribed washing steps, the complexes are selectively eluted using competitive analytes or low pH or reducing buffers for in-gel or western blot analysis.

3.4.3 Yeast two-hybrid

Yeast two-hybrid allows the identification of pairwise PPIs in vivo, in which the two proteins are tested for biophysically direct interaction. The Y2H is based on the functional reconstitution of the yeast transcription factor Gal4 and subsequent activation of a selective reporter such as His3 or LacZ.

Transcription activators are generally composed of two or more independent domains. The most basic ones are

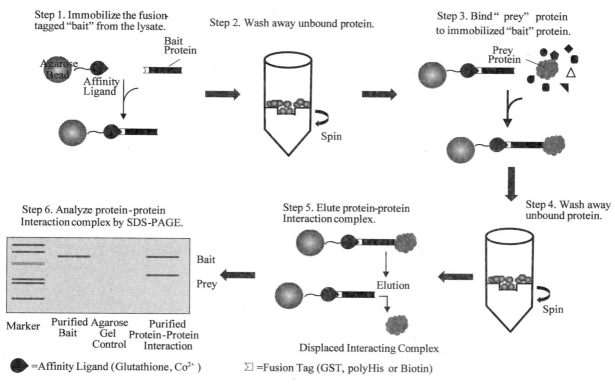

Fig. 3-2 General schematic of a pull-down assay

the DNA binding domain (BD) and the transcription activation domain (AD). Although BD alone can bind to the promoter, it cannot activate transcription. AD alone cannot activate transcription because it cannot access the promoter. Only BD and AD are spatially close to each other can they present complete transcription factor activity and activate the expression of downstream genes.

To test two proteins for interaction, two protein expression constructs are made: one protein (X) is fused to the Gal4 DNA-binding domain (DB) as the bait (DB-X), and a second protein (Y) is fused to the Gal4 activation domain (AD) as the prey (AD-Y). In the assay, yeast cells are transformed with these two constructs. Transcription of reporter genes (His3 or LacZ) does not occur unless bait (DB-X) and prey (AD-Y) interact with each other and form a functional Gal4 transcription factor. Thus, the interaction between proteins can be inferred by the presence of the products resultant of the reporter gene expression (Fig. 3-3).

Despite its usefulness, the yeast two-hybrid system has limitations. It uses yeast as main host system, which can be a problem when studying proteins that contain mammalian-specific post-translational modifications.

3.4.4 Intracellular co-localization analysis

Many biochemical methods can only be carried out in lysed cell fluid, which cannot dynamically observe the protein-protein interaction in real time under the physiological conditions of living cells. Some methods that rely on physics measurement have changed this state to a certain extent. There only briefly introduces co-localization fluorescent staining technique and fluorescence resonance energy transfer (FRET) technology.

3.4.4.1 Co-localization fluorescent staining technique

Cell imminofluorescence staining technique is a technique that combines immunology, cell biology and microscopy. Using antibodies labelled with different fluorescent groups to bind with corresponding proteins in cells and obtaining fluorescent images by confocal laser scanning microscopy, the localization information of the target

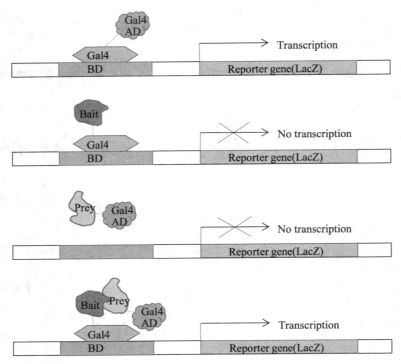

Fig. 3-3　Overview of two-hybrid assay

protein in cells can be displayed, thus providing cell-level evidence for studying the interaction between proteins. The basic process of cell immunofluorescence staining includes cell climbing, immobilization, permeabilization, blocking antibody incubation and fluorescence detection. Immunofluorescence staining technique is used to detect antigen or hapten substances in cells or tissues. According to the principle of antigen-antibody reaction, the known antibody (or antigen) is labelled with fluorescein, and then the fluorescent marker is used as a molecular probe to react with the corresponding antigen (or antibody) in tissue cells to form a specific antigen-antibody complex containing fluorescein in the cells or tissues. The fluorescein on this complex is irradiated by excitation light to emit fluorescence of various colors, and can be used for qualitative positioning and even quantitative research on antigens or antibodies in tissue cells by fluorescence microscopy observation. Proteins, polypeptides, nucleic acids, enzymes, hormones, phospholipids, polysaccharides, receptors and pathogens can be detected by immunofluorescence staining techniques. These techniques can be used for the research of co-localization of proteins (Fig. 3-4), localization of proteins on specific organelles (endoplasmic reticulum, lysosomes, etc.) and morphological studies of cytoskeleton.

3.4.4.2　Fluorescence resonance energy transfer (FRET)

FRET technique realizes continuous real-time dynamic observation of protein interactions in a single living cell. A pair of suitable fluorescent materials can form an energy donor and an energy acceptor pair, wherein the emission spectrum of the donor overlaps with the absorption spectrum of the acceptor. When they are spatially close to each other (1~10nm), the fluorescent energy generated by exciting the donor is absorbed by the nearby acceptor, so that the fluorescence intensity emitted by the donor is attenuated and the fluorescence intensity of the acceptor fluorescent molecules is enhanced. Take CFP (cyan fluorescent protein) and YFP (yellow fluorescent protein) as examples to introduce the principle of FRET (Fig. 3-5).

Fig. 3-4 Fluorescent staining results of intracellular proteins co-localization

Under the fluorescence microscope, green channel scanning showed the distribution of GFP-A in cells (left). The red channel shows the distribution of RFP-B in cells (middle). When the signals of the two proteins are merged, and appeared yellow (right), which shows that the two proteins are well co-located.

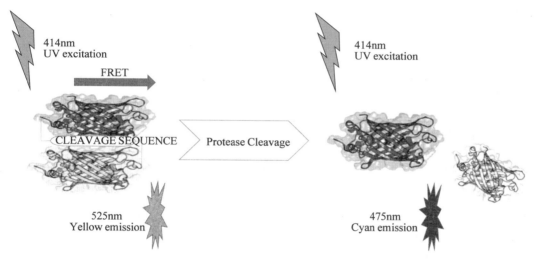

Fig. 3-5 Characteristics of fluorescent protein and basic principle of FRET

If the linker is intact, excitation at the absorbance wavelength of CFP (414nm) causes emission by YFP (525nm) due to FRET. If the linker is cleaved by a protease, FRET is abolished and emission is at the CFP wavelength (475nm).

3.4.5 Protein interaction prediction methods

PPI plays key role in predicting the protein function of target protein and drug ability of molecules. The majority of genes and proteins realize resulting phenotype functions as a set of interactions. The *in vitro* and *in vivo* methods of PPIs have their own limitations like cost, time, and so forth, and the resultant data sets are noisy and have more false positives to annotate the function of proteins. Thus, in silico methods which include sequence-based approaches, structure-based approaches, gene fusion, chromosome proximity, in silico 2 hybrid, phylogenetic profile, phylogenetic tree, mirrotree, correlated mutation and gene expression-based approaches were developed. Elucidation of protein interaction networks also contributes greatly to the analysis of signal transduction pathways. Recent developments have also led to the construction of networks having all the PPIs using computational methods for signaling pathways and protein complex identification in specific diseases. Table 3.3 lists some databases that can be used to query and predict protein interactions.

Table 3.3 **Partial protein interaction databases**

Database name	Total number of interactions (to date)	Source link
BioGrid	7,17,604	http://thebiogrid.org/
DIP	76,570	http://dip.doe-mbi.ucla.edu/dip/Main.cgi
HitPredict	2,39,584	http://hintdb.hgc.jp/htp/
MINT	2,41,458	http://mint.bio.uniroma2.it/mint/
IntAct	4,33,135	http://www.ebi.ac.uk/intact/
APID	3,22,579	http://bioinfow.dep.usal.es/apid/index.htm
BIND	>3,00,000	http://bind.ca/
PINA2.0	3,00,155	http://cbg.garvan.unsw.edu.au/pina/

3.5 Proteomics Research

The goal of proteomics is to identify, characterize, and quantify the whole content of proteins that are present in complex biological materials (tissues, cells in culture, organelles, or fluids). For the past decade, the interest for proteomic studies kept growing exponentially and today, proteomic has reached high-throughput analysis capabilities. This is the result of two major advances: ① the progress in mass spectrometry (MS) makes possible routine analysis of peptides and proteins with improved sensitivity, reliability, speed, and automation; ② the large scale genome sequence programs of the past 10 years provided large protein sequence databases for many organisms which are essential to identify quickly proteins from MS data. Mass spectrometry (MS) has become the predominant technology to analyze proteins due to its ability to identify and characterize proteins and their modifications with high sensitivity and selectivity. While mass spectrometry instruments have improved rapidly over the past couple of decades, mass spectrometry results have remained largely dependent on sample preparation and quality. Sample ionization and mass measurements are susceptible to a wide variety of interferences, including buffers, salts, polymers, and detergents. These contaminants also impair MS system performance, often requiring time consuming maintenance or costly repairs to restore function. It is important to understand the goals and expectations of the project and to choose and optimize the best sample preparation method accordingly. For example, the sample preparation requirements for protein identification from a gel slice are very different from the requirements to identify protein interaction networks, measure changes in the mitochondrial proteome, understand protein phosphorylation and signaling in cancer, or identify protein biomarkers of cancer metastasis in plasma. Unlike genomic or transcriptomic research, there is no "standard" universal sample preparation method for proteomics.

3.5.1 Protein Extraction

Tissue or cell lysis is the first step in protein extraction and solubilization. Numerous techniques have been developed to obtain the highest protein yield for different organisms, sample types, subcellular fractions, or specific proteins. Due to the diversity of tissue and cell types, both physical disruption and reagent-based methods are often required to extract cellular proteins. Physical lysis equipment, such as homogenizers, bead beaters, and sonicators, are commonly used to disrupt tissues or cells in order to extract cellular contents and shear DNA. In contrast, reagent-based methods use denaturants or detergents to lyse cells and solubilize proteins. Cell lysis also liberates proteases and other catabolic enzymes so broad-spectrum protease and phosphatase inhibitor cocktails are typically included during sample preparation to prevent nonspecific proteolysis and loss of protein phosphorylation, respectively. Different buffers, detergents and salts, cell lysis protocols can be optimized for the best protein extraction for a particular sample or protein fraction. Strong denaturants (e.g., urea or guanidine) and ionic detergents (e.g., sodium dodecyl sulfate (SDS) or deoxycholate (SDC)) solubilize membrane proteins and denature proteins. Non-ionic or zwitterionic detergents (e.g., Triton X-100, NP-40, digitonin, or CHAPS) have a lower critical micelle concentration and require lower detergent concentrations to solubilize proteins. These detergents generally solubilize membrane proteins and protein complexes with less denaturation and disruption of proteinprotein interactions.

Unfortunately, many detergents used to solubilize proteins cause significant problems during downstream mass spectrometry analysis if they are not completely removed. In addition to cell lysis buffers, detergents used to clean laboratory glassware may also contaminate samples and LC solvents. Detergents present in the sample can:

(1) Contaminate and foul autosampler needles, valves, connectors, and lines;
(2) Affect liquid chromatography by reducing column capacity and performance;
(3) Affect crystallization prior to matrix assisted laser desorption ionization (MALDI) sample analysis;
(4) Suppress electrospray ionization (ESI) prior to introduction into the mass spectrometer;
(5) Deposit in the mass spectrometer, interfering with the spectra and reducing sensitivity of the instrument.

Flexible tubing or poor quality plastic consumables can also leach phthalates and other contaminants that can interfere with downstream LC-MS analysis. Both phthalates and detergents ionize very well and overwhelm peptide signals. Polydisperse detergents, such as Triton X-100, Tween or NP-40, contain a distribution of variable length polyethylene glycol (PEG) chains that often elute throughout the LC gradient as a family of peaks separated by 44 Da mass units and overwhelm the LC-MS results. Fortunately, these leachables and detergents can often be removed by gel electrophoresis, protein precipitation, or filter-assisted sample preparation (FASP) techniques described later in this chapter.

While all detergents can affect downstream LC-MS analysis, N-octyl-beta-glucoside and octylthioglucoside are considered more compatible with mass spectrometry because they are dialyzable and monodisperse (i.e., homogeneous). In addition, a variety of mass spectrometry-compatible detergents are commercially available. Invitrosol (Thermo Scientific) contains several monodisperse detergents that elute in regions of the HPLC gradient that do not interfere with peptides or their chromatography. Cleavable detergents, such as ProteaseMax (Promega), Rapigest (Waters), PPS Silent Surfactant (Expedeon), or Progenta (Protea), degrade with heat or at low pH into products that do not interfere with LC-MS. As digestion requires incubation at 37℃ and LC-MS loading buffers contain formic acid or trifluoroacetic acid, sample preparation workflows do not require any significant modification to use these MS-compatible detergents.

3.5.2 Protein Depletion or Enrichment

Depending on the protein source and the copy number per cell, there can be a tremendous difference in the concentration between the lowest and most abundant proteins. For mammalian tissues and cell lines, protein expression can range over 6~9 orders of magnitude. For serum and plasma samples, the dynamic range can be greater than 12 orders of magnitude with serum albumin representing over 50% of the protein content. In order to get adequate depth of protein coverage in serum, to identify relevant biomarkers, abundant protein depletion is required. Although affinity chromatography using Cibacron blue dye can be used to remove albumin, immunoaffinity using antibodies is typically required to remove other abundant proteins such as immunoglobulins, transferrin, fibrinogen, and apo-lipoproteins. One advantage of using antibodies for immunodepletion is that one sample preparation technique can be used to remove the top 2~20 most abundant proteins depending on the product used. Another is that the depletion resins can be regenerated for multiple uses; though this can affect protein depletion reproducibility over time.

Protein enrichment techniques are commonly overlooked during protein sample preparation but may be necessary in order to identify and quantify biologically relevant proteins which are typically in lower abundance. One method of protein enrichment is subcellular fractionation, which separates proteins by location in a particular cellular compartment or organelle. Subcellular fractionation using sucrose density gradient centrifugation can separate vesicles and organelles including the nucleus, mitochondria, or chloroplasts from cytosolic and vesicle proteins. Differential extraction is another subcellular fractionation technique which uses detergents to selectively solubilize nuclear, chromatin-bound, membrane, cytosolic, and cytoskeletal proteins. Another method of protein enrichment is through protein modifications. Cell surface proteins which are glycosylated can be enriched by chemical labelling of oxidized glycans, metabolic incorporation of azide-containing sugars, or lectin affinity. Phosphoproteins can be enriched with immobilized metal affinity chromatography. Activity-based chemical probes are another method for enrichment of enzyme subclasses such as kinases, hydrolases, and oxidases. Finally, affinity capture using immunoprecipitation is the method of choice for enrichment of specific protein targets or protein complexes as this technique provides the highest selectivity and sensitivity for the lowest abundant proteins.

3.5.3 Protein Separation

A proteome contains a myriad of proteins. The individual proteins are separated and identified to understand the nature of different proteins and their structural and functional relationships. Proteins are isolated by a variety of means because they have a diverse set of properties. A multidimensional approach is more suitable to separate them than a single approach. Such a multidimensional approach must address the problems concerning their resolution, high throughput, automation, and adaptability to analysis by mass spectrometry. One of the most important ways they are separated is by electrophoresis, including the two-dimensional gel electrophoresis 2-DE and capillary electrophoresis. The other one is Liquid Chromatography, which is more compatible with MS analysis.

Unfortunately, many protein extraction, fractionation, enrichment and depletion methods introduce salts, buffers, detergents, and other contaminants which are not MS compatible. Because of the relative difference in molecular weight, it is simplest and preferable to remove these small molecule contaminants before protein digestion. There are a variety of options to remove these small molecules, including gel electrophoresis, chromatography, dialysis, buffer exchange, size exclusion, and protein precipitation. Gel electrophoresis is an inexpensive, straightforward method for the removal of salts, detergents, and other small molecules prior to in-gel digestion. However, keratins from skin and dust are common contaminants which can be introduced when pouring and handling gels so it is imperative to always wear gloves and to use MS grade reagents to minimize this contamination.

Reverse phase C4 or C8 cartridges can remove salts from proteins but concentrate non-ionic detergents and may have poor recovery of hydrophilic proteins. Strong cation exchange resins can remove anionic detergents, like deoxycholate or sodium dodecyl sulfate (SDS), but typically require salts for protein elution, which then have to be removed before LC-MS analysis. Dialysis membranes and cassettes are available with a variety of molecular weight cut-offs (MWCO) and can effectively exchange buffer components to remove contaminants; but dialysis is relatively slow, requires multiple buffer changes, and may be difficult with small volumes. Spin columns or stirred-cell pressure devices with MWCO membranes can rapidly exchange buffers to remove small molecule contaminants and concentrate samples. These MWCO devices allow sequential buffer exchange steps to be performed and can be used for complete MS sample preparation in the filter assisted sample preparation (FASP) methods. Size exclusion resins retain small molecules in porous beads while excluding proteins enabling rapid and efficient buffer exchange with minimal sample loss, especially in a spin column format. Notably, of all of the desalting methods available, precipitation with organic solvents such as acetone or methanol/chloroform with or without organic acids (e.g., TCA or TFA) is the most common method for desalting proteins prior to MS sample preparation as it the least expensive, simplest and most scalable option.

3.5.4 Protein Digestion

Trypsin is the most commonly used protease for MS sample preparation because of its high activity, selectivity and relatively low cost. Trypsin cleaves proteins to generate peptides with a lysine or arginine residue at the carboxy terminus. These basic amino acids at the end of every tryptic peptide improve peptide ionization and MS/MS fragmentation for peptide identification. Although trypsin is the most popular enzyme used for protein digestion, some protein sequences are not efficiently cleaved by trypsin or do not contain basic amino acids spaced close or far enough apart to generate peptides which can be used for protein identification. Trypsin digestion is less efficient at lysine and arginine residues followed by proline, repeated basic residues (e.g., KK, RK), or in the presence of post translational modifications (e.g., methylation, acetylation), resulting in missed cleavages. Some tryptic peptides may be too small to retain on reversed phase LC columns or are not unique for a particular protein. Others may be too large and hydrophobic to identify by LC-MS. For example, 56% of the tryptic peptides in yeast are $\leqslant 6$ amino acids long, while 97% of peptides identified by LC-MS are 7~35 amino acids. These short or extremely long unidentified peptides result in incomplete protein sequence coverage, resulting in missing specific peptide sequences or sites of posttranslational modifications.

For more comprehensive proteome coverage, alternative proteases are often used to generate different peptide sequences that may not be identified from tryptic digests. Partial digestion specific or non-selective proteases, like elastase or proteinase K, have been used to increase protein sequence coverage; but these proteases also increase the complexity and variability of digestion, making it more difficult to reproducibly identify the same peptides and proteins in replicate samples. Proteases with distinct cleavage specificities, such as ArgC, AspN, chymotrypsin, GluC, LysC, or LysN, produce complementary sequence information which can be combined to improve sequence coverage. This multi-enzyme approach has been used successfully by multiple laboratories to increase the number of protein identifications >10%~15% and improve the average sequence coverage by 60%~160%. Different proteases have also been shown to provide a unique repertoire of phosphopeptides which are not observed in tryptic digests. Therefore, a multiple enzyme strategy is recommended for comprehensive analysis of single proteins or complex proteomes.

3.5.5 Peptide Preparation

Once the proteins in a complex sample are solubilized, there are three general approaches to prepare protein digests: in-gel digestion, in-solution digestion, and filter-assisted sample preparation (Fig. 3-6). All three of these methods remove contaminating detergents and other small molecules, reduce and alkylate proteins, digest proteins to peptides, and prepare peptides for mass spectrometry analysis.

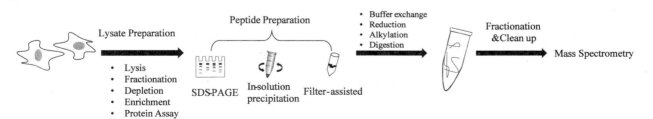

Fig. 3-6 General protein sample preparation workflow

There are many options for the extraction of proteins from tissue and cell lysates, protein fractionation and enrichment, and digestion to peptides for MS analysis.

3.5.5.1 In-Gel digestion

Sodium dodecyl sulfate-polyacrylamide gel electrophoresis (SDS-PAGE) is the most common technique for protein analysis. Gel electrophoresis is a simple, inexpensive and relatively high resolution protein separation method that can be employed in either one dimension (1D) to resolve proteins by molecular weight or two dimensions (2D) to resolve proteins by isoelectric point and molecular weight. Although 2D PAGE is not compatible with salts and ionic detergents, 1D SDS-PAGE can easily remove these and other substances which may interfere with LC-MS analysis. In fact, many academic proteomic core labs prefer or require samples to be provided in gels or gel slices because this method is so effective for sample clean up. Depending on the depth of analysis, a single band can be excised or a complex sample can then be excised as a set of gel slices in a method often referred to as GeLC-MS. Another advantage of gel-based fractionation methods is that they can reduce sample complexity and separate highly abundant proteins from lower abundant proteins. Since all of the peptides from the respective protein(s) are contained in a single gel band, spot or fraction, protein sequence coverage and posttranslational modification mapping are also improved.

After gel electrophoresis, separated proteins are detected and visualized with a variety of gel stains, including Coomassie Blue, Colloidal Coomassie, and glutaraldehyde-free silver stain. Gel bands containing protein(s) of interest are then excised, destained, reduced, and alkylated to improve digestion and peptide extraction. Disulfide bonds prevent complete protein unfolding and limit proteolytic digestion. Peptides that remain linked by disulfides are also difficult to identify due to the complexity of the peptide fragment ion spectra. Protein disulfides are typically reduced with either dithiothreitol (DTT) or tris 2-carboxyethylphosphine (TCEP) in the presence of other denaturants (i.e., heat, SDS, urea, guanidine, etc.). Reduced cysteines are then alkylated with iodoacetamide, iodoacetic acid, chloroacetamide, 4-vinyl pyridine, or N-ethyl maleimide (NEM) to prevent oxidation. Haloacetyl-containing alkylating agents are light sensitive and must be made fresh. Alkylation reactions should be performed at pH8.0 to avoid alkylation at other amino acids, and excess reagent should be quenched with DTT to prevent side reactions and over-alkylation of proteins. After reduction and alkylation, gel bands are digested with a protease; and the peptides are extracted using standard techniques. While in-gel digestion is more prone to incomplete or less reproducible digestion and lower recovery of peptides relative to in-solution option (50% ~ 70% recovery), gel electrophoresis remains an important sample preparation technique prior to MS analysis.

3.5.5.2 In-solution digestion

In-solution digestion is a popular alternative to in-gel digestion, because it requires fewer steps and can be scaled for the analysis of samples containing less than 10μg or greater than 1mg of protein. For this method, proteins are first denatured with detergents and heat or with urea or guanidine chaotropes. Disulfide bonds between cysteine residues are reduced and alkylated and then sample contaminants are typically removed by precipitation prior to digestion and cleanup. As stated above, urea has been used for many years but is not recommended because it must be made fresh as the formation of isocyanic acid over time increases the likelihood of protein carbamylation. Protein solubilization and denaturation with SDS or SDC is more effective than urea, and these detergents permit heating during the reduction of disulfides improving protein denaturation before digestion.

Once disulfides have been reduced and alkylated, contaminating salts, reducing and alkylating reagents, detergents, and small molecule metabolites present in the sample matrix should be removed from the sample before digestion. Depending on the sample source and extraction technique, small molecule contaminants may include excess protein labelling reagents, lipids, nucleotides, and phosphoryl-or amine-containing metabolites (e.g., phosphocholine, aminoglycans, etc.) that could interfere with downstream peptide enrichment or chemical tagging. These contaminants can be removed by buffer exchange using gel filtration resins, dialysis, gel electrophoresis, filtration with a molecular weight cutoff filter, or most commonly by precipitation with an acid or an organic solvent. Polydisperse detergents must be removed prior to digestion in order to prevent downstream contamination of LC-MS equipment. Most detergents can be removed by protein precipitation with four volumes of cold (-20℃) acetone. Precipitation with dilute deoxycholate and trichloroacetic acid, methanol, a 4:1:3 ratio of methanol: chloroform: water, followed by an additional three volumes of methanol, or partitioning with ethyl acetate are alternative methods of detergent removal. As an alternative workflow, digestion can be performed in 0.1% SDS or SDC, and these detergents may be removed from the peptides after digestion using a detergent removal spin column or by acidification to precipitate SDC.

Detergents, chaotropes, and organic solvent additives improve trypsin digestion efficiency and dramatically increase peptide and protein identifications in complex protein mixtures. For tryptic digestion, the protein is dissolved in a buffered solution at pH8.0 (e.g., 50~100mM ammonium bicarbonate), and digestion is performed for 4~16h at 37℃ with agitation. Low concentrations of acetonitrile, urea, SDS, SDC, or MS-compatible detergents may be included to solubilize the precipitated protein pellets and partially denature the protein to improve digestion efficiency. Endoproteinase LysC is an enzyme which cleaves after lysines similar to trypsin. Unlike trypsin, LysC can cleave at lysine residues followed by proline and is active under denaturing conditions (e.g., 8M urea). LysC digestion is often performed for 1~4h before tryptic digestion for more complete and reproducible digestion. After digestion, peptides may be desalted off-line using reverse phase solid phase extraction cartridges, tips, or on-line using a trap column before MS analysis.

3.5.5.3 Filter-assisted sample preparation (FASP)

Molecular weight cutoff (MWCO) filters have been used for decades to concentrate and exchange buffers for protein samples. FASP utilizes SDS, heat, and urea to solubilize and denature proteins before transfer to a MWCO spin column which is used for protein collection, concentration, and digestion. An advantage of FASP is that detergents, salts, and small molecules can be easily removed through multiple rounds of washing. Concentrated proteins are then alkylated, washed and digested on the membrane before elution and desalting. FASP is compatible with a wide variety of samples and has been applied to 0.2~200μg protein samples in a wide variety of applications, including brain tissue samples, formalin fixed paraffin embedded slices, C. elegans, phosphoproteomic, and glycoproteomic samples. Recently some proposed enhancements to the FASP protocol have been reported including: ① simultaneous reduction and alkylation to eliminate several centrifugation steps and improve alkylation specificity; ② prior passivation of the MWCO membrane with Tween-20 for higher peptide recovery; ③ the replacement of urea with deoxycholate for improved tryptic digestion.

3.5.6 Mass spectrometry (MS)

Fundamentally, MS measures the mass-to-charge ratio (m/z) of gas-phase ions. Mass spectrometers consist of an ion source that converts analyte molecules into gas-phase ions, a mass analyzer that separates ionized analytes based on m/z ratio, and a detector that records the number of ions at each m/z value. The development of electrospray ionization (ESI) and matrix-assisted laser desorption/ionization (MALDI), the two soft ionization techniques capable of ionizing peptides or proteins, revolutionized protein analysis using MS. The mass analyzer is central to MS technology. For proteomics research, four types of mass analyzers are commonly used: quadrupole (Q), ion trap (quadrupole ion trap, QIT; linear ion trap, LIT or LTQ), time-of-flight (TOF) mass analyzer, and Fourier-transform ion cyclotron resonance (FTICR) mass analyzer. They vary in their physical principles and analytical performance. "Hybrid" instruments have been designed to combine the capabilities of different mass analyzers and include the Q-Q-Q, Q-Q-LIT, Q-TOF, TOF-TOF, and LTQ-FTICR.

Three main methodologies are routinely used for protein identification: peptide mass fingerprinting (PMF), peptide fragment fingerprinting (PFF), and de novo sequencing.

In the case of PMF, the m/z ratio of each peptide obtained after enzymatic digestion of a protein is measured with the highest possible accuracy. The measured masses are then compared with the theoretical masses of all the peptides, which has been obtained after in silico proteolytic digestion of a selected protein database (calculated fingerprints). The degree of confidence in protein identification with this approach will strongly depend on the tight correlation between measured and theoretical masses. Therefore, the most important specification of the instrument best suited for that approach is the accuracy of mass measurement.

In the PFF approach, peptides are fragmented using a "two-dimensional" mass spectrometer (MS/MS). Intact peptide ions are selected by a first analyzer (MS1) and then dissociated by collisions, usually by passing through a neutral gas (collision-induced dissociation, CID). This results in the fragmentation of the parent peptide, which occurs at specific bonds of the polypeptide backbone. Fragment masses obtained experimentally are compared with the theoretical masses of all the fragments, which have been obtained after in silico proteolytic digestion and fragmentation of a selected protein database (calculated fingerprints). The complexity of the digestion peptide mixture will be important for the choice of the instrument and its tuning. Samples of reduced complexity are obtained when slices cut from one- or two-dimensional polyacrylamide gels are digested. When the total protein extract from the biological sample is digested and directly analyzed by MS (e.g., in shotgun proteomics), the peptide mixture is extremely complex and scanning parameters will have to be optimized. In this approach, the specifications of the best suited mass spectrometer must include: ① a collision cell generating a large number of ionized fragments; ② high accuracy of mass measurements.

These two first strategies require that the exact sequences of the studied proteins are present in the protein databases and require specialized search engines (Mascot, Sequest). If the protein database for the studied organism does not contain enough information for the comparison of fragmentation fingerprints, an alternative consists in using the so-called de novo sequencing approach. In this case, sequence information is deduced directly from the experimental MS/MS spectra by manual or automatic interpretation of the data. When a sequence of a few amino acids is obtained from an MS/MS spectrum, it can be used in a classical BLAST search to identify the protein(s). For this strategy, the same instrument specifications as the ones for PFF are required, but the highest possible accuracy in MS2 mass measurements is needed.

3.6 Research Technology of Enzymology

Substance metabolism is the basis of cell life activities. Both the energy demand of cell activities and the synthetic raw materials of biomacromolecules come from the specific metabolic pathways of glucose, lipids and amino acids in cells, which are linked and transformed with each other.

Enzymes are highly effective and extremely specific biochemical catalysts produced by living organisms and accelerate the metabolic reactions under the chemically mild conditions in the cell. Enzymes are typically proteins and their catalytic activity depends on the precise conformational structure in the folded polypeptide chains. Even minor alterations in this structure may result in the loss of enzyme activity. The catalytic properties of an enzyme may also depend on the presence of non-peptide cofactors or coenzymes. It is of practical significance to study the properties of enzyme in medicine. The study of enzymes is called enzymology. The detection of enzyme content and activity, enzymatic analysis of metabolites, isoenzyme determination are used for medical research, clinical diagnosis and curative effect judgment.

3.6.1 Separation and purification of enzymes

The general steps for enzyme purification are the same as those for common protein, including cell disruption, centrifugal separation, dialysis, precipitation, chromatography, electrophoresis and freeze drying. Attention should be paid to maintaining enzyme activity during extraction and purification such as low temperature conditions (0~4℃).

3.6.2 Determination of enzyme content

The enzyme content is regulated by gene expression, so enzyme content can be detected by routine measurement of protein. The SDS-PAGE in combination with a protein stain is widely used in biochemistry for the quick and exact separation and subsequent analysis of enzymes. Additionally, SDS-PAGE is used in combination with the Western blot for the determination of the presence of a specific enzyme in a mixture of proteins or for the analysis of post-translational modifications (i.e., phosphorylation modification). Post-translational modifications of enzymes can lead to a different relative mobility (i.e., a band shift) or to a change in the binding of a detection antibody used in the Western blot (i.e., a band disappears or appears).

3.6.3 Determination of enzyme activity

The enzymes of cellular metabolism are located within the tissue cells and are present there at high concentrations. When tissues break down or cell membranes leak, the level of these enzymes in the plasma rises. Measurement of their quantity of plasma may provide the qualitative or quantitative indexes of tissue damage. The quantity of enzyme present in a sample is usually referred to enzyme activity rather than weight or volume of an enzyme. Further determination of the enzyme activity concentration and specific activity of the enzyme is required. Enzyme activity is defined as the ability of an enzyme to catalyze a specific reaction.

Measuring enzymatic activity is vital for the study of enzyme kinetics and enzyme inhibition. Enzyme activity is a measure of the quantity of active enzyme present and is thus dependent on specified conditions. Normally, the velocity of enzyme-catalyzed reaction is proportional to the enzyme activity. The faster the reaction velocity is, the higher the enzyme activity is. Therefore, enzymatic activity is determined by the speed of a biochemical reaction catalyzed by an enzyme, and can be expressed by the consumption of substrates or the increase of products per unit time under the condition of excess substrate and most suitable other conditions. The assays of measuring enzyme activity can be split into two methods: ①Continuous assays, where the assay gives a continuous reading of activity. During the linear phase of the enzymatic reaction, the content changes of the product or substrate were measured at regular intervals; ② discontinuous assays, where samples are taken, the reaction stopped and then the concentration of substrates/products determined. It is worth noting that the initial rate of the enzymatic reaction is proportional to the activity of the enzyme to be tested (see Chapter 8).

3.6.4 Assay of isoenzyme zymogram

Isoenzymes (or isozymes) are enzymes composed of different amino acid sequences that catalyze the same chemical reaction. Isoenzymes have different kinetic parameters and different electrophoretic mobility. Tissues usually contain characteristic isoenzymes or mixtures of isozymes, that is to say, enzymes differ from one tissue to another. When enzymes are released by damaged tissue, the isoenzyme zymogram of tissue is reflected in that of serum. Therefore, knowing which isozyme is elevated in serum can be indicative of specific tissue damage. For example, creatine kinase (CK) contains two subunits. Each subunit may be either of the muscle (M) or the brain (B) type, and there are three isozymes (MM, MB, and BB) found in different tissues. CK-MB makes up about 25% of myocardium, but it is not found in any other tissues, so CK-MB is a significant marker for a myocardial infarction.

3.6.5 Study of enzyme kinetics

The enzyme kinetics is aimed to study the rate of enzyme-catalyzed reaction and the factors that affect the velocity, including substrate concentration, enzyme concentration, temperature, pH, inhibitor, activator, and so on (see Chapter 8). The study on enzyme kinetics can reveal the catalytic mechanism of enzyme, its role in metabolism, how its activity is controlled, and how a drug might inhibit or activate the enzyme.

3.6.6 Enzymatic analysis

Enzymatic analysis is an analytic method using enzymes as tools, based on which the quantity of substrates, coenzymes, activators or inhibitors can be measured. Since the enzyme method has good specificity, simple and convenient operation, mild reaction, no pollution, and high sensitivity (10~100mol/L), it has been widely used in clinical chemistry determination of most substances in body fluids.

The commonly used methods include kinetic assay, end point assay and ELISA. The end point assay is the most common method in which sample is incubated with the buffered substrate for a fixed period of time and the amount of substrate used or product formed can be quantitatively measured when the reaction is stopped. End point assay can be divided into single enzymatic reaction assay and coupled enzymatic reaction assay (see Chapter 8).

Section II Basic Biochemical Techniques

Chapter 4 Basic Operation of Biochemical Experiments

Biochemistry and molecular biology continues to be an essential part of modern biological research. It is of practical significance to study the structure and functions of genes and proteins in medicine, therefore, setting up a usual laboratory and grasping the basic operation of biochemical and molecular biology experiments are necessary.

For setting up a laboratory of protein study, we should consider to prepare some basic supporting materials, such as glassware and plasticware, frequently-used chemicals and disposables, small equipment and accessories, equipment and apparatus, at the same time, detection and assay requirements, spectrophotometer, electrophoretic apparatus, pH meter, microcentrifugator, water baths, etc. are necessity in the lab. The number and types of fractionation requirements are available and outstanding features for a protein purification laboratory. The usefulness of such a centrifuge is directly related to the presence in the laboratory of a wide variety of rotors and centrifuge tubes and bottles.

The necessity for maintaining a stable pH when studying proteins or enzymes is well established. We will discuss these items in detail, some basic operations will be mentioned here only briefly.

4.1 Solution Preparation

There are many factors that must be considered when choosing a buffer. When studying an enzyme one must consider the pH optimum of the enzyme, nonspecific buffer effects on the enzyme, and interactions with substrates or metals.

According to the purity of the chemical reagent, it can be divided into four grades according to the impurity content: the first grade reagent is high quality pure reagent, usually expressed by GR; the second grade reagent is analytical pure reagent, usually expressed by AR; the third grade reagent is chemical pure reagent, usually expressed by CR; the fourth grade reagent is experimental or industrial reagent, usually expressed by LR.

In addition, there are some special purity standards, such as spectral purity, fluorescence purity and so on. Different specifications of reagents should be selected according to different experimental requirements.

4.1.1 The procedure of solution preparation

The preparation steps of the reagent include calculation, weighing, dissolution, pipetting, washing beaker, adjusting pH value, constant volume and shaking.

Examples: prepare 1L of 0.5mol/L EDTA solution (pH8.0) steps:

(1) Calculate the mass of solute $EDTA \cdot 2Na \cdot 2H_2O$ required: 0.5mol/L×1L×372.2g/mol=186.1g

(2) Weighing: 186.1g solid $EDTA \cdot 2Na \cdot 2H_2O$ on balance.

(3) Dissolution: stir 186.1g $EDTA \cdot 2Na \cdot 2H_2O$ dissolved in a beaker with distilled water of 800mL.

(4) Transfer the solution into the measuring cylinder. Note that the measuring cylinder is marked as the standard volume at a certain temperature. The solution must be cooled before it can be transferred to the measuring cylinder.

(5) Wash the beaker with distilled water and add it into the measuring cylinder.

(6) Add NaOH to the side agitation and adjust the pH value. Note that the solute must be completely dissolved.

(7) Add distilled water to constant volume; shake and shake well and transfer to reagent bottle for storage or subpackage for autoclave sterilization and storage.

Measuring cylinder usage and points for attention:

(1) The solute is dissolved in the beaker and then moved into the measuring cylinder.

(2) Can not be heated, can not replace the reagent bottle to store the solution.

4.1.2 Solution storage

Some common solutions can be prepared into a high concentration of storage solution (e.g., 0.5M Tris-Cl; 1M IPTG; 10×loading buffer, etc.), according to the storage conditions stored in room temperature, refrigerator, pay attention to avoid light, sealing, etc. When used, the volume ratio should be diluted with water, or the new solution can be added in a small volume. Store at low temperature and restore the liquid properties of the solution again (e.g., 10% SDS, 1M DTT, etc.).

4.2 Pipettes Operation

4.2.1 Graduated Pipette

The graduated pipette is usually applied to transport a measured volume of liquid with higher accuracy (Fig. 4-1). There are two types: complete drain-out and incomplete drain-out.

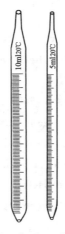

Fig. 4-1 Graduated pipette

For the complete drain-out pipettes, the liquid on the tips which cannot automatically flow out is included in the measured volume, it should be blown out and "Blow" is labelled on their outside walls. However, as the liquid on the tips of the incomplete type is not included in the measured volume, you should lean the tips against the inside walls of the containers and wait for several seconds until no liquid flow out rather than blow it out.

The graduated pipettes also have A-grade and B-grade types which can be chosen according to the accuracy. The common volumes are 0.1~10mL.

During the operation, use a rubber bulb to fill the pipette, and keep the graduation to the operator. It is proper to immerse the tip under the surface of the liquid level about 1cm. If immersed too deep, more liquid will attach to the outside walls making a bigger error. However, if immersed too shallow, air might be inhaled into the pipettes or liquid may go into the bulbs. After the liquid reaching the required graduation line on the pipette, remove the tip from the liquid and discard excess liquid vertically until the lowest part of its meniscus is level with the required graduation line. Then transfer the liquid to the received container and drain out.

The size of the pipette should be selected correctly as per the required volume to avoid the artificial expansion of the error range. For example, taking 0.1mL of the solution using 1mL pipette will increase the value of error, or taking 1mL of the solution using 0.5mL pipette twice rather than 1mL pipette once will also increase the value of error.

4.2.2 Adjustable Micropipettor

The adjustable micropipettor is commonly applied to accurately transport the solution which volume is less than 1mL, consisting of the plastic rod enclosure, graduation knob and top button (Fig. 4-2). There is a disposable plastic tip corresponds to type each of micropipettors. It has high accuracy and could be operated easily. Briefly, it is an accurate piston which can adjust the volume continuously by changing the route of the piston, that is, the volume transported is the multiplication of the moving distance and the sectional area of the piston.

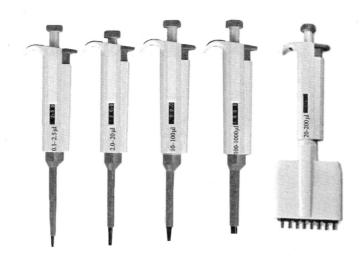

Fig. 4-2 Adjustable micropipettors

The adjustable micropipettors come in $2\mu L$, $10\mu L$, $20\mu L$, $50\mu L$, $100\mu L$, $200\mu L$ and $1,000\mu L$, up to $5,000\mu L$ volumes, which should be selected properly depending on the requirement to avoid the artifitial expansion of the errors and the influence on the results.

A complete operation includes five steps: volume setting, tip installation, liquid suction, dispensing the liquid, and tip discard. These steps have their respective operating rules.

1. Volume setting

Set the desired volume by turning volume adjustor clockwise to increase volume or counterclockwise to decrease volume.

2. Tip installation

Place a proper-size tip on the lower end of the pipettor.

3. Liquid suction

Draw the liquid into the tip by slowly releasing the plunger.

4. Dispensing the liquid

Transfer the pipet to the receiving container and depress the button past the first stop to the second stop, this dispenses the liquid and blows out the last drop to insure accuracy.

5. Tip discard

Remove tip by pressing down on the tip discarder. Remember to change tips between solutions to avoid mixing or contaminating the solutions used.

4.3　Centrifuge Operation

A machine running on a centrifugal force to separate different parts is called a centrifuge. Centrifugal force is the body in the high-speed rotation by the power. They are commonly used to separate components of different densities, such as water and solids.

4.3.1　Microcentrifuge

Accelerate your micro-volume separations with our compact, safe, and easy-to-use microcentrifuges, offering power and versatility for routine laboratory work as well as high-end applications.

Find the ideal blend of capabilities to support your micro-volume protocols—from basic microtube processing to nucleic acid mini-preps and spin-columns to hematocrit capillaries—in a small footprint with:

Intuitive controls, easy-to-read displays, and fast one-click centrifuge lid closure.

Fast acceleration and deceleration, and outstanding corrosion resistance with lightweight engineered polymer rotors.

Wide variety of rotors support 0.2mL, 0.5mL, 1.5mL, 2.0mL microtubes, PCR strip tubes, 5mL tubes, and hematocrit.

Capillary tubes.

Fig. 4-3　Microcentrifuge

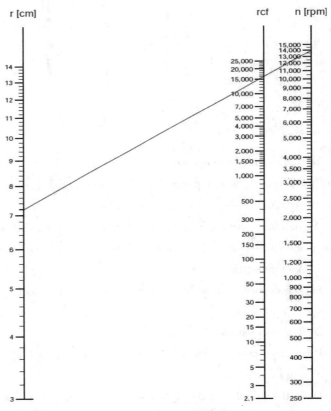

Fig. 4-4　Conversion of centrifugal force in centrifuges

1. Usage

Sample balance, open power switch, parameter setting, start operation, stop operation, close power supply, clean centrifugal chamber and rotor, close machine cover.

2. Notes

(1) Centrifuge tubes must be placed symmetrically before centrifugation.

(2) Before centrifuge starts, there must be no sundries in the centrifuge room.

(3) When working in centrifuge, the setting speed should be lower than the maximum speed of the rotor.

(4) In the whole process of centrifugation, the machine cover must be shut down until the centrifuge stops rotating.

Chapter 5　Spectrophotometry

Spectrometry is a qualitative and quantitative technology based on emission spectra, absorption spectra or scattering spectra of the material. Spectrometry can be classified in many different types depended on the form of spectra, including atomic emission spectrometry, flame photometry and fluorescence spectrometry based on the character of emission spectra, UV and visible spectrophotometry based on the absorption spectra, atomic absorption and infrared spectroscopy based on scattering spectral character.

Spectrophotometry is based on the reflection or transmission properties of a substance. It has some advantages, such as simple, rapid, high sensitivity, accuracy, and selectivity. So it is one of the most widely used analytical techniques in biochemistry laboratory and clinical research.

5.1　Basic Concepts and Principle of Spectrophotometry

Light is a kind of electromagnetic wave. The light can be seen by our naked eye is really a very small portion of the electromagnetic spectrum (Fig. 5-1) within the wavelength range between approximately 380nm and 760nm. The visible light could be split into many colors by a prism. Each color is signed by the wavelength of light. Any solution that contains a subtance that absorbs the visible light will appear a color. Light with wavelengths longer than 760nm or shorter than 380nm is invisible. Ultraviolet is a region of the electromagnetic spectrum that has a wavelength range from 12.5nm to 380nm. Ultraviolet-visible spectrophotometry using lights in the visible range and adjacent near ultraviolet (UV) range (200~380nm) is discussed in this chapter.

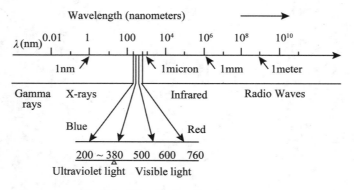

Fig. 5-1　The electromagnetic spectrum

5.1.1　Absorption spectrum

Each substance has its own characteristic spectrum. The light absorption of the same substance at different wavelengths is different, and light absorption of different substances at the same wavelength is also different. Spectrophotometry is based on this selective absorption of light by a substance. Because the extent to which a sample absorbs light depends upon the wavelength of light, spectrophotometry is performed using monochromatic light.

To clearly describe the selective absorption of light by a substance, an absorption spectrum is usually plotted, which shows how the absorption of light varies with the wavelength of the light (Fig. 5-2). The extent of light absorption is commonly referred to as absorbance (A) or extinction (E). The absorption spectrum is a plot of absorbance vs wavelength and is characterized by the wavelength at which the absorbance is the greatest (λ_{max}). The λ_{max} is the characteristic of each substance and provides information on the electronic structure of an analyte. Unknown substance can be identified by their characteristic absorption spectrum and λ_{max} (qualitative analysis).

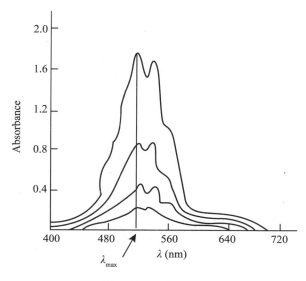

Fig. 5-2 Absorption spectrum

5.1.2 Basic laws of absorption of light——Lambert-Beer's law

The quantitative analysis of spectrophotometry is based on the basic laws of absorption of light—Lambert-Beer's law. When monochromatic light (light of a specific wavelength) passes through a solution there is usually a quantitative relationship between the solute concentration and the absorption of the light, which is described by the Lambert-Beer's law. Therefore the concentration of solute can be measured by determining the extent of absorption of light at the appropriate wavelength. In order to obtain the highest sensitivity and to minimize deviations in quantitative measurement, spectrophotometric measurements are usually made using light with a wavelength of λ_{max}.

For a uniform absorbing medium the portion of the light passing through it called the transmittance (T). $T = I/I_0$, where I_0 is the intensity of the incident light, I is the intensity of transmitted light. The absorbance of the light is equal to the logarithm of the reciprocal of the transmittance: $A = \lg(1/T)$.

According to Lambert-Beer's law, when a ray of monochromatic light passes through an absorbing solution, absorbance of the solution is directly proportional to the concentration of the absorbing substance and the depth of the solution through which the light passes. Equation for Lambert-Beer's law is: $A = \lg(1/T) = KCL$, where K is absorption coefficient (the proportionality constant that depends on the absorbing substance, wavelength of light and the temperature), C is the concentration of the substance absorbing light, L is the length of the path of the light. A is dimensionless. When length L is in centimeter and concentration $C = 1\text{mol/L}$, the absorbance is equal to "ε" (molar extinction coefficient), which is written as $\varepsilon^{1\text{mol/L}}$ and has a dimension of $1\text{mol}/(\text{L}\cdot\text{cm})$. Since the molar extinction coefficient may be very large, an alternative is $E^{1\%}$, which represents the extinction given by 1cm thick sample of a 1% solution of the substance. If the $\varepsilon^{1\text{mol/L}}$ or $E^{1\%}$ is known, the amount of the substance can be quantified, $C = A/\varepsilon$, or $C = A/E^{1\%}$.

If the Lambert-Beer's law is obeyed, a plot of absorbance against concentration gives a straight line passing through the origin. Sometimes, a non-linear plot is obtained of absorbance against concentration. This is probably not satisfied with one or more of the following prerequisites of Lambert-Beer's law.

(1) Light must be of a narrow wavelength range, preferably monochromatic.

(2) The wavelength of light used should be λ_{max}.

(3) The solution should be stable and uniform. There must not be ionization, association, dissociation of the solute during the process of spectrophotometric measurement.

(4) Solution concentration is not too high ($A = 0.15 \sim 0.7$).

5.2 Measurement of Light Absorption

5.2.1 Components of spectrophotometer

A spectrophotometer is employed to measure the light absorption of a sample. There are many kinds of spectrophotometers. All the spectrophotometers employ the basic components illustrated in Fig. 5-3.

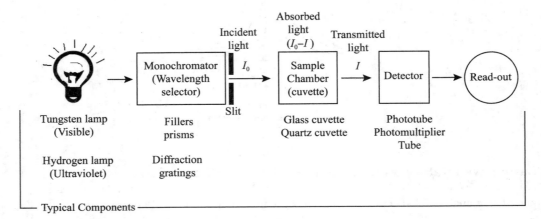

Fig. 5-3 The main components of a spectrophotometer

5.2.1.1 Light sources

The source of visible light is tungsten lamp that emits light in the range of 340 ~ 900nm. Some spectrophotometers employ an additional deuterium lamp emitting light in the range of 200 ~ 360nm for spectral analysis in the UV range.

5.2.1.2 Wavelength selector

1. Monochromator

A monochromator is an optical device that transmits a mechanically selectable narrow band of wavelengths of light. Selection of wavelengths may be done by the use of optical filters, prisms of quartz, or diffraction grating. However, it is not feasible to have a light of a single wavelength. The monochromatic light produced by monochromator is a light that has its maximum emission at a specific wavelength, with progressively less energy at longer and shorter wavelengths. Therefore, the purer the monochromatic light is, the sensitive the measurement is.

2. Slit system

The spectrophotometer places narrow slit in the light path that confine the light beam to a narrow path and also help to exclude light from extraneous sources. Then the intensity of the incident light can be adjusted.

5.2.1.3 Sample container—cuvettes

Glass cuvette is for visible light measurement. Quartz cuvette is for UV measurement (below 340nm). Absorbance characteristics of cuvettes should always be checked in order to obtain a standardized set for spectrophotometric measurement.

5.2.1.4 Detector systems

Vacuum phototubes and photomultiplier tube are usually used to detect the transmitted light. Most detectors have a scale that reads both in absorbance units, which is a logarithmic scale, and in % transmittance, which is an arithmetic scale. The absorbance scale is normally read directly for calculation of sample concentration.

5.2.2 Using a spectrophotometer

Turn on the spectrophotometer and allow 15min for warm up of the instrument prior to use. Use the wavelength knob to set the desired wavelength. With the sample cover closed, use the zero control to adjust the meter recorder to "0" on the % transmittance scale. Insert a clean cuvette containing the blank solution into the sample holder. The amount of solution placed in the cuvette is usually about 2/3 of the total volume of the cuvette. Close the cover and use the light control knob to set the meter recorder to "0" on the absorbance scale (100% transmission). Remove the blank cuvette, insert a cuvette holding the sample solution and close the cover. Read and record the absorbance. Remove the sample tube, readjust the zero, and recalibrate if necessary before checking the next sample. Carefully clean and store cuvettes for later use.

5.2.3 Requirements of spectrophotometric measurement

When use a spectrophotometric measurements, one must understand that the absorption is produced by the particular absorbing substances (specific absorbance), but the solvent and substances in the reagents (nonspecific absorbance). The assay must include the following solutions:

1. Blank solution/reference solution

A proper blank solution contains none of the assayed substance, but all other chemicals in the test or standard solution and undergoes the same stages as the standard and test solution. This solution will help to exclude the absorption due to reagents (nonspecific absorbance).

2. Standard solution

It contains all the reagents of test and blank but also includes a solution of known concentration of the substance which is going to be determined in the test solution. It helps to correlate the absorption with the concentration of the concerned substance.

3. Test solution/ sample solution/ determined solution

It contains all the reagents present in the blank and standard and undergoes the same steps, but an unknown quantity of the concerned substance.

5.3 Applications of Spectrophotometry

5.3.1 Qualitative analysis

Spectrophotometry can be a very useful technique for identifying unknown compounds. Absorption spectra of a pure unknown compound and a standard known compound can be generated by measuring the absorbance at a variety of wavelengths. The shape of the spectra, λ_{max} (wavelength of maximal absorption) and ε (molar extinction coefficient) can be compared to identify the property of the unknown compound. If the two spectra are completely consistent, the sample and standard compounds can be tentatively determined to have the same chromophore group, and may be the same compound. Some substances with the same chromophore group but different molecular structure may also generate the same absorption spectra, but their absorption coefficient is different. If the absorption spectra, λ_{max} and ε are complete same, they are the same material. However, for really precise qualitative analysis other assays are required.

5.3.2 Quantitative analysis (determining the unknown concentration)

5.3.2.1 Standard addition method

The experimental approach exploits Lambert-Beer's Law, $A = KLC$, which predicts a linear relationship between the absorbance of the solution and the concentration of the analyte on the condition that K and L are constant. Prepare the test solution, the standard solution and the blank solution. All the solutions undergo the same treatment and their absorbances are measured under the same experimental conditions. Since the same absorbing substance, wavelength of light, temperature and length of the path of light, the K and L of the test and standard solutions are same. Let the concentration of the test unknown solution and standard solution is C_u and C_s, and the absorbance of unknown solution and standard solution is A_u and A_s, respectivly. $A_u = KLC_u$ and $A_s = KLC_s$, then $KLC_u/KLC_s = A_u/A_s$. So, C_u can be calculated by the formula as following:

$$C_u = \frac{A_u}{A_s} \cdot C_s$$

5.3.2.2 Calibration curve method

Calibration curve (also called working curve) shows how absorbance changes with the concentration of a solution. The curves are constructed by measuring the absorbance from a series of standards of known concentration and used not only to determine the concentration of an unknown sample but also to calibrate the linearity of an analytical instrument. A series of standard solutions are prepared in calibration curve method then the absorbances are measured and used to prepare a calibration curve (standard curve), which is a plot of absorbance *vs* concentration. The absorbance of the unknown solution is detected and then used to determine the concentration of the compound in the test solution in conjunction with the calibration curve (Fig. 5-4).

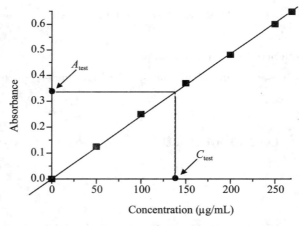

Fig. 5-4 Detection the concentration of an unknown solution by calibration curve

As a perfect calibration curve, assays should normally be performed in duplicate at least while preparing standard curve and only the mean should be plotted. There must be at least five points while plotting calibration curve. The points on the calibration curve should yield a straight line. The best straight line should be drawn through the points, the origin or other points. The calibration curve may vary in different batches of reagents and hence calibration curve will be done in each. Calibration curves should never be extrapolated beyond the highest absorbance value measured. If A is outside the range of the curve, the sample should be diluted or concentrated.

Experiment 1 Quantitative Analysis of Protein

Protein is the biological macromolecule which plays important roles in the cell. Protein quantitative analysis can not only help us to determine the protein content of samples but also support the diagnosis of human diseases. Proteins are composed of many amino acids linked by peptide bond. Amino acid residues, peptide bond, and some characteristics of physical and chemical properties of protein are the basis of different methods of quantitative analysis of proteins. In samples, there are many types of protein. In order to determine the quantity of the concerned protein, we need to base on the different physical and chemical properties of proteins separately. This chapter only focused on the quantitative determination of total protein in sample.

The Kjeldahl method of nitrogen analysis is the worldwide standard for calculating the protein content in a wide variety of samples, which is an analytical method based on the characteristic of element composition of proteins. The nitrogen content of protein is approximately 16% of the total weight, namely, 1g nitrogen content is equal to 6.25g protein. Nitrogen in the sample is converted into ammonia by digestion process. The ammonia is distillated from the digestate and collected. Then the amount of ammonia is quantified by titration and the initial protein concentration is then calculated.

Protein samples often consist of a complex mixture of many different proteins. The quantitative measurement is usually achieved on the basis of the dye-forming reactions between functional group of the protein assayed and dye-forming reagent. The intensity of the dye correlates directly with the protein concentration and can be measured by colorimetric determination. The colorimetric protein determination methods include Folin-Phenol (Lowry) assay, BCA assay, Bradford assay and Biuret assay.

The tyrosine and tryptophan residues of proteins exhibit an ultraviolet absorbance at approximately 280nm and the concentration of protein in a pure solution is generally proportional to the absorbance at 280nm, so UV spectrophotometry can also be used to measure protein concentration.

In general, there is no completely satisfactory single method to determine the concentration of protein in any given sample. The choice of the method is based on the purpose, the nature and amount of protein in sample, and the nature of the other components in the sample. The characteristics of the commonly used methods are summarized in the Table 5.1.

Table 5.1 **Methods of protein concentration determination**

Method	Sensitivity range (μg/mL)	Variation between different protein	Compatibility with other detergents	Characteristics
The Kjeldahl method		little		accurate, complicated, low reproducibility
UV spectro-photometry	100~1,000	large	low	rapid, simple, recoverable
Biuret assay	1,000~10,000	little	low	high reproducibility
Folin-Phenol (Lowry) assay	25~250	large	low	convenient and inexpensive
Bradford assay	10~1,000	large	low	fast, simple
BCA assay (bicinchoninic acid)	10~1,000	large	high	sensitive and rapid, broad linear working range

Protocol I Determination of Protein Concentration by Lowry Method

Principle

Under alkaline conditions, copper complexes with protein. When Folin-Phenol reagent (phospho-molybdic-phosphotungstic reagent) is added, it binds to the protein. Bound reagent is slowly reduced and changes color from yellow to blue.

The chromogenic principle of Folin-Phenol reagent includes the following two reactions:

(1) Peptide bonds in proteins + Cu^{2+} in alkaline copper sulfate reagent $\xrightarrow{OH^-}$ Protein-Cu^{2+} compounds

(2) Residues of tyrosine and tryptophan in protein-Cu^{2+} compounds + Phosphomolybdic acid-phosphotungstic acid in phenol reagent $\xrightarrow{reduction}$ bluish-green compounds (molybdenum/tungsten blue)

The intensity of the blue-green color of the Folin-Phenol reaction is proportional to protein concentration. Therefore, Lowry Method, also called Folin-Phenol method, can be used to quantify proteins.

This method is simple and very sensitive. Its detectable range is 25~250μg/mL protein. It is more subject to interference by a wide range of non-proteins, additives such as EDTA, NP-40, Triton X-100, barbital, CHAPS, cesium chloride, citrate, cysteine, diethanolamine, dithiothreitol, HEPES, mercaptoethanol, phenol, polyvinyl pyrrolidone, sodium deoxycholate, sodium salicylate and Tris. There is also much protein-to-protein variation in the intensity of color development because of different compositions of tyrosine and tryptophan in different proteins. The standard protein should be similar to the unknown. For serum, use bovine serum albumin as a standard since albumin is a major component of serum.

Reagents

1. Alkaline copper sulfate reagent:

Solution A: Sodium carbonate (Na_2CO_3) 10g, sodium hydroxide (NaOH) 2g and potassium/sodium tartrate (potassium saline or sodium saline) 0.25g are dissolved in 500mL distilled water (dH_2O).

Solution B: Cupric sulfate ($CuSO_4 \cdot 5H_2O$) 0.5g is dissolved in 100mL dH_2O.

Solution A 50mL and Solution B 1mL are mixed before using. The mixture is alkaline copper sulfate reagent. This reagent must be used within 24h.

2. Phenol reagent: Sodium wolframate ($Na_2WO_4 \cdot 2H_2O$) 100g and sodium molybdate 25g are added into 1.5L flask with ground and round bottom and are dissolved in 700mL dH_2O. Then, 85% phosphoric acid (H_3PO_4) 50mL and concentrated hydrochloric acid (HCl) are added and mixed completely. Thereafter, the ground condensation tube is joined, reflux at small fire for 10h. After refluxing and cooling, the condensation tube is taken off. Then, lithum sulfate ($Li_2SO_4 \cdot H_2O$) 150g, dH_2O 50mL and several drops of bromine water are added. Continuing boiled for 15min with the flask mouth opening in order to get rid of excessive bromine (in draft chamber). Cooling again, the solution becomes yellow or gold. (If it's still green, several drops of bromine water must be added again.) The solution is diluted to 1L and filtrated. The filtrate is stocked in brown bottle, which is the stock solution. The acidity of the stock solution is roughly 2N. It is standardized by titrating standard NaOH (about 1N), and phenolphthalein is an indicator. It is the end point when the color of the solutions turns from red to purplish red, purplish gray, and suddenly green.

Phenol reagent applied solution is made when the stock solution is diluted by equal volume of dH_2O.

3. Protein standard solution (250μg/mL): dissolve 25mg of crystalline bovine serum albumin in 100mL of 0.9% NaCl.

4. 0.9% NaCl solution.

5. Sample: serum.

Procedure

1. Prepare standard curve and determine serum sample.

(1) One blank solution, five standard solutions and one sample solution in duplicate are set up and operated according to the following table.

Serum (0.1mL) is accurately taken to volumetric flask (50mL). Then 0.9% NaCl is added up to the graduation, mixed completely. Then put 0.5mL diluted serum into the sample tube, which is processed simultaneously with standard tubes.

Reagents (mL)	Blank	1	2	3	4	5	Sample
Protein standard solution	—	0.1	0.2	0.3	0.4	0.5	—
Diluted serum	—	—	—	—	—	—	0.5
0.9% NaCl	0.5	0.4	0.3	0.2	0.1	—	—
Alkaline copper sulfate reagent	2.5	2.5	2.5	2.5	2.5	2.5	2.5
Mix and incubate 20min at room temperature							
Phenol reagent	0.25	0.25	0.25	0.25	0.25	0.25	0.25
Equal protein concentration (μg/mL)	0	50	100	150	200	250	?

(2) Each tube must be mixed immediately after phenol reagent is added (Attention: Cloudiness shouldn't appear, otherwise, chromogenic degree would be weakened). After phenol reagent is added into the tubes for 30min, absorbance is read at wavelength 650nm on condition that blank tube is adjusted to zero.

2. Quantitative analysis of serum proteins.

(1) Plot a standard curve of absorbance versus protein concentration of standards. According to absorbance of sample, concentration of diluted serum can be looked up in the standard curve. The concentration of original serum can be further determined by multiplying times of dilution.

(2) Serum protein concentration can also be calculated according to the following formula:

$$\text{Serum protein concentration (g/L)} = \frac{A_{sample}}{A_{standard}} \cdot C_{standard} \cdot \text{Times of dilution}$$

Protocol II Determination of Protein Concentration by BCA Method

Principle

The principle of bicinchoninic acid (BCA) method is similar to Lowry method, in which Cu^{2+} is first reduced to Cu^+ forming a complex with protein amide bonds. Then the Cu^+ reacts with BCA reagent to form a stable purple color compound, which is detectable at 562nm. The absorbance value is proportional to protein concentration. Therefore, BCA method can be used to quantify proteins.

This assay has the advantage that it can be carried out as a one-step process compared to the two steps needed in the Lowry assay. Using a 96-well plate detected by a spectrophotometric microplate reader in the experiment makes the assay quicker and easier and requires less sample volume and less BCA reagents.

Reagents

1. BCA reagents:

Reagent A: 1g sodium bicinchoninate (BCA), 2g sodium carbonate, 0.16g sodium tartrate, 0.4g NaOH, and 0.95g sodium bicarbonate, brought to 100mL with distilled water. Adjust the pH to 11.25 with 10M NaOH.

Reagent B: 0.4g Cupric Sulfate · $5H_2O$, in 10mL distilled water.

Standard working solution: Mix reagent A and reagent B according to volume ratio of 50 : 1. The working solution is stable for 1 week and should be green.

2. BSA protein standard solution (1mg/mL): dissolve 10mg of bovine serum albumin (BSA) powder in 10mL of 0.9% NaCl.

3. 0.9% NaCl solution.

4. Sample: serum. Prepare serial dilutions of serum with 0.9% NaCl solution.

Procedure

1. Make a dilution series of BSA protein standard solution to prepare a set of protein standards of known concentration (100μg/mL, 200μg/mL, 400μg/mL, 600μg/mL, 1000μg/mL).

2. Pipette 20μL of each standard or diluted sample in duplicate (or triplicate) wells in a microplate, including blanks lacking protein.

3. Add 200μL working reagent to each well.

4. Mix well by pipetting up and down several times. Do not introduce any bubbles into the solution.

5. Incubate at 37℃ for 30min. Cool samples before reading the absorbance.

6. Determine the absorbance at 562nm (A_{562}) using a spectrophotometric microplate reader. Subtract the average blank from the measurements of all other standards and samples.

7. Calculate concentration of serum according two ways:

(1) Plot a standard curve of A_{562} versus protein concentration of standards. According to A_{562} of sample, concentration of diluted serum can be looked up in the standard curve. The concentration of original serum can be further determined by multiplying times of dilution.

(2) Serum protein concentration can also be calculated according to the following formula:

$$\text{Serum protein concentration (g/L)} = \frac{A_{sample}}{A_{standard}} \cdot C_{standard} \cdot \text{Times of dilution}$$

Protocol III Determination of Protein Concentration by Bradford Method

Principle

Bradford method also called coomassie brilliant dye-binding method. In acidic solution, coomassie brilliant blue (CBB) G250 could combine with protein, and then it shows a shift in its absorption maximum from 465nm to 595nm, and its colour changes from red to blue. The absorption at 595nm is directly proportional to protein concentration in a definite range of 1 to 1,000μg/mL. So we can determine protein concentration by this method.

This method is easy to perform, quick, highly sensitive, stable and has less interfering substances. It is a widely used method, especially for determining protein content of cell fractions and samples for gel electrophoresis.

Reagents

1. 0.9% NaCl solution.

2. 100μg/mL protein standard solution: dissolved 10mg bovine serum albumin in 100mL 0.9% NaCl solution.

3. Coomassie brilliant blue G250 solution: dissolve 0.1g Coomassie brilliant blue G250 in 50mL ethanol (95%), then add 100mL 85% (m/v) phosphoric acid. Dilute to 1L when the dye has completely dissolved and filter through Whatman #1 paper just before use. This solution can be preserved for 1 month at room temperature.

4. Sample solution: serum. Prepare serial dilutions of serum with 0.9% NaCl solution.

Procedure

1. Prepare standard curve and determine serum sample.

One blank solution, five standard solutions and one sample solution in duplicate are set up and operated according to the following table:

Reagents (mL)	Blank	1	2	3	4	5	Sample
Protein standard solution	—	0.2	0.4	0.6	0.8	1.0	—
Diluted serum	—	—	—	—	—	—	1.0
0.9% NaCl solution	1.0	0.8	0.6	0.4	0.2	—	—
Coomassie bright blue G250	5.0	5.0	5.0	5.0	5.0	5.0	5.0
Equal protein concentration ($\mu g/mL$)	0	20	40	60	80	100	?

Mix the solution and allow the tubes to stand at room temperature for 10min. Determine absorbance at 595nm on condition that blank tube is adjusted to zero. The used cuvettes should be immediately washed by 95% alcohol after the measurement.

2. Calculate concentration of sample.

(1) Plot a standard curve of absorbance versus protein concentration of standards. According to absorbance of sample, concentration of sample can be looked up in the standard curve.

(2) Serum protein concentration can also be calculated according to the following formula:

$$\text{Serum protein concentration (g/L)} = \frac{A_{\text{sample}}}{A_{\text{standard}}} \cdot C_{\text{standard}} \cdot \text{Times of dilution}$$

Protocol IV Determination of Protein Concentration by Biuret Method

Principle

Under alkaline conditions, the compounds containing two or more peptide bond (—CONH—) react with copper sulfate in Cu^{2+} forming a purple complex which can be measured at 540nm. This reaction is called biuret reaction. Protein contains many peptide bonds and so it can be determined based on the biuret reaction.

Biuret method is fast, convenience and has very few interfering agents such as ammonium salts. It is the most widely used method for total protein determination. However, this assay has a relatively low sensitivity and a low specificity (biuret reagent could also react with the compounds containing groups such as —$CONH_2$、—CH_2NH_2、—$CS-NH_2$).

Reagents

1. 6mol/L NaOH: Dissolve 60g NaOH in 250mL distilled water.

2. Biuret reagent: Dissolve $CuSO_4 \cdot 5H_2O$ 2.5g, potassium tartrate 10.0g and potassium iodide 5.0g, and added in 100mL 6mol/L NaOH, constant volume to 1L. Store in brown container. Discard if a black precipitate forms.

3. 0.9% NaCl solution.

4. 10mg/mL protein standard solution: dissolved 500mg bovine serum albumin (BSA) in 50mL 0.9% NaCl solution.

5. Sample solution: serum. Prepare serial dilutions of serum with 0.9% NaCl solution.

Procedure

1. Prepare standard curve and determine serum sample.

One blank solution, five standard solutions and one sample solution in duplicate are set up and operated according to the following table:

Reagents (mL)	Blank	1	2	3	4	5	Sample
Protein standard solution	—	0.2	0.4	0.6	0.8	1.0	—
Diluted serum	—	—	—	—	—	—	1.0
0.9% NaCl solution	1.0	0.8	0.6	0.4	0.2	—	—
Biuret reagent	4.0	4.0	4.0	4.0	4.0	4.0	4.0
Equal protein concentration (mg/mL)	0	2	4	6	8	10	?

Mix the solution and allow the tubes to stand at room temperature for 30min. Determine absorbance at 540nm on condition that blank tube is adjusted to zero.

2. Calculate concentration of sample.

(1) Plot a standard curve of absorbance versus protein concentration of standards. According to absorbance of sample, concentration of sample can be looked up in the standard curve.

(2) Serum protein concentration can also be calculated according to the following formula:

$$\text{Serum protein concentration (g/L)} = \frac{A_{sample}}{A_{standard}} \cdot C_{standard} \cdot \text{Times of dilution}$$

Protocol V Determination of Protein Concentration by UV spectrophotometry

Principle

The ultraviolet absorption characteristic of proteins depends on the Tyr and Trp content (and to a very small extent on the amount of Phe and disulfide bonds). The absorption value at 280nm is directly proportional to the protein content. Different proteins have widely varying content of Tyr and Trp, therefore the A_{280} varies greatly between different proteins. Unlike the colorimetric process, this method is less sensitive and requires higher protein concentrations and is suitable for concentration determination of pure protein solutions when a standard protein with similar content is available.

The advantages of this method are that it is simple, and the sample is recoverable. But A_{280} may be disturbed by the parallel absorption of non-proteins (e.g., DNA). The maximal ultraviolet absorption of nucleic acids is at 260nm. Therefore, through measuring the absorption at 280nm and 260nm, it is possible to correct for the nucleic acid present and roughly estimate protein concentration in a sample using an empirical formula. However, the absorbance at 260nm and 280nm are variable in different proteins and nucleic acids that errors may be caused in measurement.

The specific extinction coefficient of a number of proteins at 280nm has been determined, so the amount of the pure samples can also be quantified using the known $\varepsilon^{1mol/L}$ or $E^{1\%}$.

Reagents

1. Distilled water (dH_2O).

2. 10mg/mL protein standard solution: dissolved 500mg bovine serum albumin (BSA) in 50mL 0.9% NaCl solution.

3. Sample: protein solution with unknown concentration.

Procedure

1. Mix 1mL of standard or sample solution with 2mL of dH_2O and measure the absorbances (A) at wavelength 280nm and 260nm respectively using quartz cuvette with 1cm light path, on condition that absorbance of dH_2O is adjusted to zero.

2. Quantitative analysis of sample.

(1) Read A_{280} and A_{260} of the sample, then calculate protein concentration using the following formula: Lowry-Kalckar formula:

$$\text{Protein concentration (mg/mL)} = 1.45A_{280} - 0.74A_{260}$$

Warburg-Christian formula:

$$\text{Protein concentration (mg/mL)} = 1.55A_{280} - 0.76A_{260}$$

(2) If the sample is a pure protein with known $\varepsilon^{1mol/L}$ or $E^{1\%}$ at 280nm, the amount of the substance can be quantified using the following formula:

$$C = \frac{A}{\varepsilon} \quad \text{or} \quad C = \frac{A}{E^{1\%}}$$

Concentration unit is g/dL, or mol/L, depending on which type of extinction coefficient is used.

Experiment title

Date
Observations and results

Discussion

1. The most important application of spectrophotometry in biochemistry lab and clinical research is _____.

2. _____ is a plot which show absorbance changes with the concentration of a solution.

3. Absorption at 280nm by proteins depends on the _____ and _____ content.

4. How to prepare a perfect calibration curve?

5. What are the colorimetric methods used to determine the concentration of serum protein? Describe the basic principle of each method briefly.

Teacher's remarks
Signature
Date

Experiment 2 Determination of Blood Glucose Level Regulated by Hormones

Part I Modified Ortho-Toluidine in Boric Acid method (O-TB method)

Principle

Glucose is hexose containing an aldehyde group. Heated up under the acidic condition, it dehydrates to be 5-hydroxyl-methyl-2-furfural, which condenses with ortho-toluidine to produce a stable green compound. Its color intensity is directly proportional to glucose content of the sample. Glucose contents of samples can be determined by compared with standard glucose solution assayed by the same method.

Reagents

1. Saturated boric acid solution: 6g boric acid is dissolved in 100mL dH$_2$O, place it overnight and then filter it.
2. O-Toluidine reagent: 1.5g thiourea is dissolved in 883.2mL acetic acid and mixed with 76.8mL or tho-toluidine. Then add 40mL saturated boric acid solution into it. The reagents may be stored in brown reagent bottle for several months.
3. Stock glucose standard solution: 10.0g anhydrous glucose is dissolved in 1L 0.25% benzoin acid.
4. Working glucose standard solution: 1.0mg/mL working glucose standard solution is prepared by ten times dilution of glucose standard stock solution with 0.25% benzoin acid.

Procedure

1. Three test tubes are used and the operation is done according to the following table:

Reagent (mL)	Blank tube	Standard tube	Test tube
dH$_2$O	0.1	0.0	0.0
Blood plasma	0.0	0.0	0.1
Glucose standard solution	0.0	0.1	0.0
O-Toluidine reagent	5.0	5.0	5.0

2. After mixing the tubes, heat the tubes at 100℃ (boiling water) for 5~6min and then cool them in running water. The absorbance is read within 30min at wavelength 620nm under the condition that absorbance is adjusted to zero with the blank tube. Samples with a very high concentration will produce a very high absorbance which can not be read accurately on a spectrophotometer. Their absorbances can be measured again after being diluted with O-Toluidine reagent.

3. Blood glucose concentration can also be calculated according to the following formula:

$$\text{Blood glucose levels (mg/dL)} = \frac{A_T}{A_S} \cdot C_S \cdot 100$$

Clinical significance

1. Glucose is a major sugar in blood. Blood glucose levels fluctuate physiologically as well as in diseased condition. Therefore, glucose measurement is a very important assay. It is most commonly performed in the

detection and treatment evaluation of diabetes.

2. Normal blood glucose level in fasting serum/ plasma is 70~110mL/dL or 3.89~6.11mmol/L. A condition in which an excessive amount of glucose circulates in the blood is referred as hyperglycemia. And hypoglycemia refers to a pathologic state produced by a lower than normal blood glucose level.

3. The concentrations of glucose will increase in some patients, who suffer from endocrine disease such as DM (diabetes mellitus), endocrine diseases (hyperthyroidism, hypercorticism). Moderate rise of glucose level is associated with infectious diseases, intercranial diseases like meningitis, encephalitis, tumors and hemorrhage. Anesthesia also causes increase in glucose level. Physiological hyperglycemia occurs due to the take-up of high sugar food, or the increased secretion by adrenal gland because of tight mood.

4. Hypoglycemia is due to overdose of insulin or oral hypoglycemic agents or a variety of hormonal causes. Low levels of glucose under fasting conditions are associated with hypothyrodism, hypoadrenalism, hypopituitarism and also in glycogen storage disease. The concentration will be low in some long-term malnutrition people or the men who can not eat. Physiological hypoglycemia may occur due to starvation or violent exercise.

Part II Effects of insulin and adrenalin on blood glucose level

Principle

Blood glucose levels in human are mainly regulated by hormones. Insulin reduces blood glucose but the other hormones such as adrenalin can increase blood glucose. In this experiment, insulin is injected into one rabbit and adrenalin is injected into another. The venous blood is collected from the two rabbits before and after injecting and the blood glucose levels of the samples are determined. Analyze the variations of blood sugar before and after injection and learn the effects of insulin and adrenalin on the concentration of blood glucose.

Reagents

1. Potassium oxalate.
2. 25% glucose.
3. Adrenalin.
4. Insulin.
5. Dimethylbenzene.

Procedure

1. Prepare experimental animals: select two normal rabbits, take their weights after fasting overnight.

2. Collect blood samples before injection: collect the blood from ear-border vein. Hairs around ear-border vein are firstly removed. Clean the skin and make the vein hyperaemic with dimethylbenzene. Then the vein is poked with a blade. Venous blood is collected into blood-collection tubes containing 2mg/mL final concentration of potassium oxalate and the tubes are shaked while collecting to prevent the blood from coagulating. Stop bleeding with cotton by pressing blood vessel. Finally, isolate plasma from the collected whole blood.

3. Inject hormones: one rabbit is subcutaneously injected insulin at a dosage of 0.75units/kg body weight. The other one is injected adrenalin at a dosage of 0.4mg/kg body weight. The time of injection are recorded respectively.

4. Collect blood samples after injection: follow the same operations as in step 2. The time for blood collection: 30min after adrenalin injection and one hour after insulin injection. After blood collection, inject subcutaneously 10mL 25% glucose to prevent the rabbit from insulin shock (hypoglycemia shock).

5. Measure blood glucose level of the samples: the method is the same as in Part I.

6. Analyze the effects of hormones: calculate the increased and decreased percentage of blood glucose after the injection of insulin and adrenaline respectively.

Clinical significance

1. Insulin decreases glucose level.

Insulin is the only hormone to reduce the blood glucose level. It enhances the uptake of glucose into the liver, stimulates glycogen synthesis and glycolysis, and inhibits gluconeogenesis.

2. Adrenalin increases glucose level.

Adrenaline is a stress hormone, which is secreted as a result of stressful stimuli (fear, excitement, hemorrhage, hypoxia, hypoglycemia, etc.) and leads to glycogenolysis in liver and muscle. It blocks insulin secretions, impairs insulin action in target tissues, so that hepatic glucose production is increased and the capacity to dispose exogenous glucose is impaired.

Experiment title

Date

Observations and results

Discussion

1. Normal blood glucose level in fasting plasma is _____ mL/dL or _____ mmol/L.
2. Which hormones elevate or reduce blood glucose?
3. Why does the blood glucose level increase as a result of stressful stimuli?

Teacher's remarks
Signature
Date

Chapter 6 Electrophoresis

Electrophoresis, one kind of electrokinetic phenomenon first discovered by Reǔss in 1809, has been developed as a typical biochemical technique and universally practiced throughout the fields of biology, biochemistry and molecular biology. Some theoretic knowledge including basic principles, impact factors, detection of separated components and applications of special electrophoresis, and common experiments of electrophoresis applying in medical laboratories are introduced in this chapter.

6.1 Basic Principles of Electrophoresis

Electrophoresis is the forced migration of charged particles in an electric field. Cations move toward the cathode and anions move toward the anode. An electrophoretic system consists of a voltage power supply, electrodes, buffer, and a supporting medium such as filter paper, cellulose acetate strips, polyacrylamide gel, or a capillary tube (Fig. 6-1). Electrophoresis is a simple, rapid, and sensitive technique to separate and purify the charged biomolecules including amino acids, proteins, and nucleic acids.

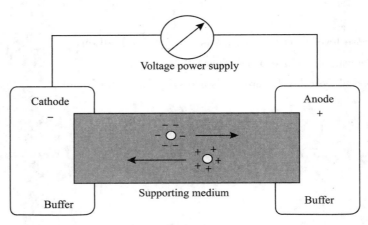

Fig. 6-1 An electrophoretic system

When an electric field is applied, a charged particle will experience a forward electric force F, and a reverse frictional force f due to frictional resistance of the matrix. The electric force is proportional to the net charge on the molecule, Q, and the electric field strength, E.

$$F = EQ$$

If electrophoresis occurred in free solution rather than within a gel, the frictional force of a globular molecule would follow Stokes' law:

$$f = 6\pi r v \eta$$

where r is the radius of a particle, η is the viscosity of the medium, and v is the migration velocity of the particle. When the two forces, F and f reach equilibrium ($EQ = 6\pi r v \eta$), the particle will migrate in the electric field with a constant velocity:

$$v = \frac{EQ}{6\pi r\eta}$$

The migration velocity is usually represented by the electrophoretic mobility, M.

$$M = \frac{v}{E} = \frac{Q}{6\pi r\eta}$$

In an electrophoretic system, molecules have different migration mobilities depending on their total charge, size and shape, and can therefore be separated after a given time.

Stokes' law, however, is not sufficient to describe the frictional force of a particle within a gel matrix. In addition to the viscosity of medium, the frictional resistance is also determined on the density and effective "pore size" of the gel. The result of this combination of factors is that among molecules with the same charge to mass ratio, larger molecules move slowly in gel electrophoresis and electrophoretic separation occurs by size. In techniques such as denaturing DNA/RNA agarose gel electrophoresis or protein SDS-polyacrylamide gel electrophoresis (SDS-PAGE), conditions are maintained so that the charge to mass ratio is virtually equal among sample molecules and separation occurs based on the molecular size.

6.2 Impact Factors of Electrophoresis

6.2.1 Properties of the molecules

6.2.1.1 Molecular charge

Sample molecules, a mixture of proteins for example, have different net charges depending on the pI of molecule and the pH of the electrophoresis buffer. When an electric field is applied, charged molecules migrate in the direction of the electrode bearing the opposite charge. Namely, the positively charged molecules migrate towards the cathode and the negatively charged molecules migrate toward the anode of an electrophoretic apparatus. The more charges they have, the faster they migrate. Hence, proteins can be separated based on their charges.

6.2.1.2 Molecular shape

Molecules can be separated by gel electrophoresis because of their difference in shape. Compared with globular proteins, the long and loose proteins tend to interact with the gel network more tightly and therefore travel more slowly.

6.2.1.3 Molecular size

The molecules with different size can be separated by electrophoresis, regardless of their same charge and shape. Of course, a smaller one has a faster electrophoretic mobility.

6.2.2 Properties of the electrophoretic system

6.2.2.1 Electric field strength

The mobility of molecules can be improved by increasing the strength of electric field. However, the electric current and heat are increased subsequently, which may cause damages of the molecules. So, an efficient cooling system is essential in a high-voltage electrophoresis.

6.2.2.2 Support medium

Common zonal support media used in electrophoresis are paper and starch, cellulose acetate membrane

(CAM), agarose gel, and polyacrylamide gel. Paper electrophoresis is one of the earliest forms of zone electrophoresis, in which a strip of filter paper is used as a medium to support a thin layer of buffer. CAM give sharper bands and better resolution than paper, bind proteins less and have less electroendosmosis. At present, CAM electrophoresis replaces paper electrophoresis and remains in use, mainly in medical diagnostic laboratories. Agarose is a polysaccharide obtained from seaweed. Agarose gel has a macroreticular structure, and zero electroendosmosis. It can be cast onto a sheet of flexible plastic gel-bond and it can be dried to provide a durable record. Polyacrylamide gel has the advantage of being a synthetic gel, which is highly reproducible. Moreover, the pore size can be controlled by varying the proportions of acrylamide and the crosslinking agent, bisacrylamide. The microreticular nature of polyacrylamide gels has sieving effect which greatly increases its resolution and sensitivity. Support media may influence separation in three aspects: restrictions on mobility, electroendosmosis and effect of molecular sieve.

1. Electroendosmosis

The support medium and/or the surface of the separation equipment, such as glass plates, tubes or capillaries can carry charged groups (e.g., carboxylic groups in starch and agarose, sulfonic groups in agarose, silicon oxide on glass surfaces). In basic and neutral buffers, these groups ionize to negatively charged ions and will be attracted by the anode when an electric field is applied. Because these negatively charged groups are immobilized, so that a counter-flow of H_3O^+ (hydronium) ions towards the cathode will happen. This effect is electroendosmosis and may result in blurred zones and distort the migration of the samples.

2. Effect of molecular sieve

Gel is a porosity support medium, which can act as a sieve to improve resolution by reducing diffusion broadening of bands. The larger the molecule is, the slower it migrates. Pore sizes in the gels can be controlled by regulating the concentration of the monomer units of the gel, depending on the samples being separated.

6.2.2.3 Properties of buffer

The pH value of the buffer influences the charge density of a molecule, and will influence the migrating direction and velocity of the molecule consequently. In vertical or capillary systems, the pH is very often set to a very high (or low) value, so that as many as possible sample molecules are negatively (or positively) charged, and thus migrate in the same direction.

Ionic Strength also influences rate of separation. The weaker is the ionic strength, the faster is the migration velocity of charged particles. However, if the ionic strength is too low, the buffering ability of the buffer will also decrease. This will increase the diffusion of samples and decrease the resolution of electrophoresis. Ionic strength (I), can be calculated from the following equation:

$$I = \frac{1}{2} \sum c_i z_i^2$$

where the sum is over all ionic species of concentration c_i and valence z_i. Normally, the most suitable ionic strength is about 0.02 to 0.2.

6.3 Detection of Components after Electrophoretic Separation

Many macromolecules can not be observed directly, different samples including proteins and nucleic acids should be stained by corresponding methods.

6.3.1 Detection of proteins

6.3.1.1 Staining by special protein dye

Chemical dyes such as amido black-10B and Coomassie brilliant blue (CBB) R250 or G250 are commonly used for staining and visualization of proteins. A more sensitive staining method, silver staining is commonly used for detecting proteins in polyacrylamide gel electrophoresis.

6.3.1.2 Detection by fluorescence labelling

Fluorescent dyes (Cy2, Cy3 and Cy5) can be used to bind the amino groups of proteins, which are detected by analysis of the relative fluorescence intensities.

6.3.1.3 Detected by radiolabelling

Isotopes ^{14}C, ^{3}H, and ^{32}P are commonly used to label proteins, which can be then detected with an autoradiographic analysis.

6.3.1.4 Immunochemical detection

In Western blotting, the protein on the membrane can be recognized and bound with its specific antibody, and the corresponding second antibody, which is conjugated with an enzyme such as horseradish peroxidase (HRP). This is visualized by a colorimetric reaction catalyzed by HRP. Immunochemical detection is usually used to identify a specific protein in a mixture.

6.3.2 Enzymes

Appropriate enzymatic methods will be used, in which the products or substrates of the catalyzed reaction can be detected by an instrument.

6.3.3 Detection of glycoproteins

Schiffs Reagent and Alcian blue are used for staining and detecting neutral glycoprotein and acidic mucopolysaccharides, respectively.

6.3.4 Lipid detection

Sudan Black is a dye commonly used to stain lipids.

6.3.5 Staining of nucleic acids

Ethidium Bromide (EB) or SYBR green are commonly used to bind DNA and visualized under UV light.

6.4 Special Electrophoretic Techniques and Applications

Based on the development of support media and basic principle of electrophoresis, many kinds of electrophoretic techniques have been generated and widely used in life science researches. Some special electrophoretic techniques are introduced as following.

6.4.1 Isoelectric focusing (IEF)

Isoelectric focusing (IEF) is an electrophoretic technique that has been widely used for the separation of proteins based on differences in their isoelectric points (pI). IEF takes place in a pH gradient, which is formed by special amphoteric buffers (ampholytes) on an electric field. When an electric field is applied, the negatively and positively charged ampholytes move towards the anode and cathode respectively, and a stable gradually increasing pH gradient depending on the initial mixture of ampholytes is formed. The ampholytes are generally provided in a support matrix, a polyacrylamide gel or an agarose gel. During the electrophoresis, proteins or peptides migrate through a stable pH gradient until they reach the pH equal to their pI, where the net charge and mobility of proteins are zero. In case of diffusion to an adjacent pH-environment, molecular particles will rapidly acquire a charge and move back again and will "focus" at their pI site.

Because pore sizes of polyacrylamide gels can be accurately controlled by regulating the total acrylamide concentration and degree of cross-linking (relationship between acrylamide and bis-acrylamide), the polyacrylamide gel is more commonly used in IEF to separate some charged macromolecules, such as proteins and isoenzymes. Commercial immobilized pH gradient (IPG) strips with different pH ranges are available, which facilitate the application of IEF. IEF is also used in some complex separation techniques to increase separation efficiency, such as two-dimensional polyacrylamide gel electrophoresis (2D-PAGE, or 2DE), in which IEF is coupled with SDS-PAGE, and capillary isoelectric focusing (CIEF), in which IEF is combined with a capillary electrophoresis.

6.4.2 Capillary electrophoresis (CE)

Capillary electrophoresis (CE) is a special electrophoretic technique that performs electrophoresis by using narrow-bore fused-silica capillaries to separate a complex array of large and small molecules. The capillary tube has a high surface to volume ratio, so that it easily radiates heats and prevents the denaturation of biomolecules including proteins and enzymes during the high-voltage electrophoresis. Applying injection of samples to the capillary tubes and output of the detection digital to a computer is of great benefit to the automation of CE.

The capillary tubes can be filled with different matrices according to different samples. Depending on the type of capillaries and electrolytes used, CE can be classified into several types. For example, capillary zone electrophoresis (CZE), which is based on differences in the charge-to-mass ratio of analytes; capillary gel electrophoresis (CGE), which is the adaptation of traditional gel electrophoresis into the capillary using polymers in solution to create a molecular sieve, and commonly used for DNA sequencing and genotyping; capillary isoelectric focusing (CIEF) that allows amphoteric molecules to be separated by electrophoresis in a pH gradient generated between the cathode and anode, and is commonly used to determine a protein's pI.

6.4.3 Two-dimensional polyacrylamide gel electrophoresis (2D-PAGE, or 2DE)

Two-dimensional polyacrylamide gel electrophoresis is a special complex electrophoresis of IEF and SDS-PAGE, named as 2D-PAGE or 2DE, and is the most popular protein separation technique in proteomics study. The separation of proteins is performed with the first dimension IEF according to their pI, and the second dimension SDS-PAGE according to their molecular weights (MW). Usually, proteins are subjected to SDS-PAGE in a perpendicular direction of IEF. After separation in first dimension of IEF using a long thin immobilized pH gradient (IPG) strip, proteins are stained as bands. The IPG strip is then inserted into the gel cassette on top of an SDS-PAGE slab gel. When the second dimension electrophoresis is finished, the protein pattern, in the gel is a number of spots rather than bands (Fig. 6-2).

At present, 2D-PAGE is a powerful tool of proteomics research to analyze differences of protein expressional profiles. Traditionally, 2D-PAGE gels are stained using Coomassie brilliant blue (CBB) or the more sensitive silver staining procedures, and analyzed using computer software for changes in protein expression. 2-dimensional difference gel electrophoresis (2D-DIGE) is a modified 2D-PAGE method, in that each sample is prelabelled with

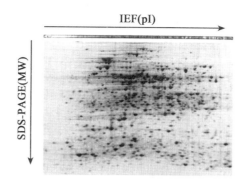

Fig. 6-2　A typical profile of 2D-PAGE

a fluorescent dye prior to IEF. Three dyes (Cy2, Cy3 and Cy5) are commercially available and therefore up to 3 samples can be run on the same gel at any one time. Analysis of 3 samples on one 2D-DIGE gel rather than three individual 2D-PAGE gels will reduce the experimental gel-to-gel variation.

6.4.4　Pulsed-field gel electrophoresis (PFGE)

Pulsed-field gel electrophoresis (PFGE) is a technique used to separate especially long strands of DNA (10~2000kb). During common agarose gel electrophoresis, the electric field is continuous and DNA fragments above 30kb migrate with the same mobility regardless of size, lead to a single large diffuse band. In PFGE, the DNA is forced to change direction during electrophoresis by pulsed alternation of electric fields, different sized fragments (within the diffuse band in common agarose gel electrophoresis) begin to separate from each other (Fig. 6-3). Specialized equipment is required to perform PFGE, which is at least consisting of a gel rig with clamped electrodes in a hexagonal design, a chiller, a pump, and a programmable power supply.

PFGE separation effect is associated with some parameters, such as the applied voltage and field strength, pulse length, the agarose concentration, the buffer chamber temperature, and the amount of DNA loaded. Usually, the longer the pulse time, the larger the DNA fragments to be suitable for separation.

PFGE has been widely used for gene mapping and medical epidemiology studies, such as epidemiological typing of pathogenic bacteria, identifying restriction fragment length polymorphisms (RFLP), detecting in vivo chromosome breakage and degradation, and determining the number and size of chromosomes.

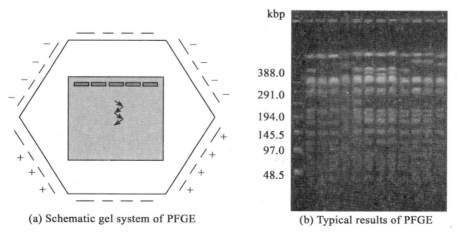

(a) Schematic gel system of PFGE　　(b) Typical results of PFGE

Fig. 6-3　The principle and results of PFGE

Experiment 3   Separation of Serum Proteins by Cellulose Acetate Membrane Electrophoresis

Principle

Separation and detection of serum/plasma proteins is a routine work in medical diagnostics. The most common method is the electrophoresis by using a strip of cellulose acetate membrane (CAM) as the supporting medium. Since the membrane served only to support the buffer, CAM electrophoresis can be considered as a form of free electrophoresis. Serum proteins are negatively charged in alkaline buffer, so they will migrate toward anode in an electric field. Because of the differences of net charge and molecular weight, their migration velocities are distinct. Proteins separated on film are fixed in position and stained with a protein-specific dye, such as amido black. After destaining, five zones indicating fractions of albumin, α_1-globulin, α_2-globulin, β-globulin, and γ-globulin of serum proteins can be visualized. Their percentages can be estimated by scanning with densitometry. A typical result is shown in Fig. 6-4.

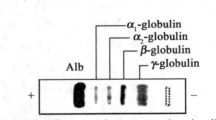
(a) Pattern of separated serum proteins visualized in the CAM strip

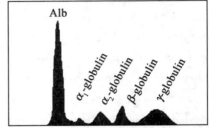
(b) Densitometric scanning of the CAM strip converts bands to characteristic peak of 5 fractions

Fig. 6-4   Seperation of serum proteins by CAM electrophoresis

Reagents

1. The electrophoresis buffer: 0.05mol/L barbital sodium / barbital (pH8.6). Take 12.76g barbital sodium and 1.66g barbital and dissolve them in 1L distilled water (dH_2O).

2. Staining solution: take 0.5g amido black 10B, dissolve it in 40mL dH_2O, 50mL methanol and 10mL glacial acetic acid.

3. Destaining solution: 45mL 95% alcohol, 5mL glacial acetic acid, and 50mL dH_2O.

Procedure

1. Immerse a 10cm × 2cm cellulose acetate membrane (CAM) strip into the electrophoresis buffer, take it out and absorb water on the membrane with filter paper.

2. Add 3~5μL sample on the rough side of membrane at the site 2cm from cathode, and put the membrane on the electrophoresis device with the rough side of membrane down, connect the electrode and the membrane with filter paper. Turn on the power and electrophoresis is carried out at a constant voltage of 10V/cm. The electrophoresis continues for 50~60min.

3. After electrophoresis, take out CAM strip, immerse into the dye solution for 5min.

4. Put the strip into the destaining solution, shaking the strip with a forceps for rinsing, and repeat three times

till the background of the membrane is colorless.

5. Observe and judge the shade of color and width of the bands by naked eyes to estimate their quantity roughly or estimate their percentages by scanning with densitometry.

Reference

The normal reference range of human serum protein components is shown in the table below.

Table 6.1 **Reference values of serum protein components**

Proteins	Concentration (g/L)	Percentage (%)
Total protein	64 ~ 83	100
Albumin	43 ~ 53	57 ~ 72
α_1-globulin	0.4 ~ 2.6	1.8 ~ 4.5
α_2-globulin	2.5 ~ 5.3	4.0 ~ 8.3
β-globulin	4.0 ~ 8.2	6.8 ~ 11.4
γ-globulin	7.6 ~ 18.6	11.2 ~ 23.0

Clinical significance

The serum proteins are actually a very complicated mixture that includes not only simple proteins but also mixed or conjugated proteins such as glycoproteins and various types of lipoproteins. The fractions of the serum proteins separated by CAM electrophoresis, especially α_1-globulin, α_2-globulin, β-globulin, and γ-globulin are mixtures. Albumin is the major protein of human serum produced by liver about 12g per day, it has two main functions: ① maintaining colloid osmotic pressure of blood; ② transportation: acting as a carrier molecule for bilirubin, fatty acids, trace elements and many drugs. The significant changes of the total and fractions of serum proteins are associated with some diseases.

1. Decreased total protein may indicate malnutrition, nephrotic syndrome, and gastrointestinal protein-losing enteropathy.

2. Increased α_1-globulin may be seen in rheumatoid arthritis, systemic lupus erythematosus (SLE), malignancy and acute inflammatory disease; decreased α_1-globulin may be seen in α-1 antitrypsin deficiency.

3. Increased α_2-globulin may appear in acute inflammation, nephrotic syndrome, and chronic inflammation; decreased α_2-globulin may appear in hemolysis.

4. Increased β-globulin may happen in myeloma, estrogen therapy, and hyperlipoproteinemia (such as familial hypercholesterolemia); decreased β-globulin may happen in congenital coagulation disorder, consumptive coagulopathy, disseminated intravascular coagulation.

5. Increased γ-globulin may be the results of multiple myeloma, rheumatoid arthritis, SLE, chronic hepatitis, hyperimmunization, and Waldenstrom's macroglobulinemia.

Experiment title

Date
Observations and results

Discussion

1. The support medium of electrophoresis influences separation in three aspects: _____, _____, _____.

2. Serum proteins were separated by cellulose acetate membrane electrophoresis into five zones: _____, _____, _____, _____, _____, respectively. These proteins in CAM electrophoresis were detected by staining with _____.

3. What are the functions of serum proteins?

Teacher's remarks
Signature
Date

Experiment 4 Separation of Serum Lipoproteins by Agarose Gel Electrophoresis

Principle

In alkaline buffer solution, serum lipoproteins are negatively charged. They will migrate toward anode in electric field. Because of the difference in their compositions, the lipoprotein particles have different charges and sizes, which lead to their different migration velocities. Serum lipoproteins can be separated into at least four kinds of bands by agarose gel electrophoresis, including α-lipoproteins, pre-β-lipoproteins, β-lipoproteins and chylomicrons (CM). The four major bands are related to the different lipoprotein layers obtained by ultracentrifugation (Fig. 6-5).

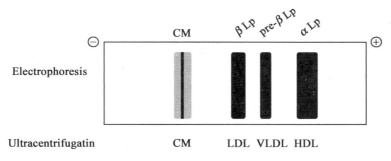

Fig. 6-5 Pattern of separated serum lipoproteins

Prestaining with a lipid dye Sudan black B, the bands of serum lipoproteins can be seen during electrophoresis. The results of percentages of lipoproteins are estimated as the shade of color and width of the bands by naked eye, or calculated by scanning with a spectrophotometer.

Reagents

1. The electrophoresis buffer: take 15.5g barbital sodium and dissolve it in distilled water, add 1mol/L HCl to adjust pH to 8.6, and make the final volume up to 1L with distilled water.

2. The agarose gel buffer: take 10.3g barbital sodium and dissolve it in distilled water, add 1mol/L HCl to adjust pH to 8.6 and make the final volume to 1L with distilled water.

3. Dye solution: dissolve 1g Sudan Black B in 100mL Petroleum ether/ethanol (1 : 4, V/V) solution.

4. 0.8% agarose gel: take 0.8g agarose powder and made the final volume up to 100mL with the agarose gel buffer in a flask. Heat the flask in a water bath and boil it for 30min with occasional agitation.

Procedure

1. Prepare an agarose gel: take a flask containing 0.8% agarose gel and heat it in a microwave oven for 2~4min until the agarose dissolves completely. Pour the warm agarose solution onto a glass slide, and insert a comb at the side 2cm from cathode for forming the sample wells in the gel. Allow 20~30min for solidification.

2. Sample preparation: mix 0.2mL serum with 0.02mL dye solution and 0.01mL ethanol. Settle the mixture at 37℃ for 30min. Centrifuge it 5min at 2000r/min.

3. Sample loading: carefully remove the comb and load 20μL sample mixture into the wells. Place the gel in a horizontal electrophoresis apparatus. Connect the electrode and the gel with filter paper. Run the electrophoresis at

a constant voltage of 120V for 45~60min.

4. Observe the shade of color and width of the bands by naked eyes to estimate their quantity, or estimate their percentages by scanning with densitometry.

Reference

Reference value ranges of fasting serum lipoproteins in normal adults:

α-lipoprotein: (25.7 ± 4.1)%;

pre-β-lipoprotein: (21 ± 4.4)%;

β-lipoprotein: (53.3 ± 5.3)%.

Clinical significance

Electrophoretic analysis of serum lipoproteins has a great clinical significance. It can be used to detect primary disorders of plasma lipoproteins such as hyperlipoproteinemia, which are the metabolic disorders that the concentrations of specific lipoproteins in fasting serum were abnormally elevated. It can be classified into five types:

Type I: CM (+)

Type IIa: β lipoprotein (LDL) ↑

Type IIb: β lipoprotein (LDL) ↑, pre-β lipoprotein (VLDL) ↑

Type III: β lipoprotein (LDL) ↑↑

Type IV: pre-β lipoprotein (VLDL) ↑

Type V: pre-β lipoprotein (VLDL) ↑, CM (+)

Experiment title

Date
Observations and results

Discussion

1. Agarose gel has a macroreticular structure, and can be obtained in a form with zero _____.

2. Serum lipoproteins can be separated into at least four kinds of bands by agarose gel electrophoresis, including _____, _____, _____ and _____. These lipoproteins can be detected by staining with _____. They are related to the different lipoprotein layers obtained by ultracentrifugation _____, _____, _____ and _____.

3. Which bands may disappear in the electrophoresis of lipoprotein in the fasting serum of a healthy subject?

Teacher's remarks
Signature
Date

Experiment 5 Proteins Separation by SDS-Polyacrylamide Gel Electrophoresis

Principle

Polyacrylamide gel electrophoresis (PAGE) is a technique that uses a gel made of polymerized acrylamide. Monomers of acrylamide, $CH_2{=}CH{-}C({=}O){-}NH_2$, are polymerized into long linear polymer, polyacrylamide, which can be cross-linked with N, N'-methylene bisacrylamide, $CH_2{=}CH{-}C({=}O){-}NH{-}CH_2{-}NH{-}C({=}O){-}CH{=}CH_2$, to form a gel matrix. Polymerization is catalyzed by free radicals, generated by agents such as ammonium persulfate (AP) in the presence of N, N, N', N'—tetramethylethylenediamine (TEMED). The gel pore size is determined by the total acrylamide concentration and degree of cross-linking that regulated by varying ratios of acrylamide to bis-acrylamide. If the gel pore size approximately matches the size of the molecules to be separated, smaller molecules can move freely in an electric field, whereas larger molecules have restricted movement. Thus, proteins separate into discrete bands based on the gel pore size and protein's charge, size and shape.

A discontinuous gel system is also used frequently in PAGE, consisting of a stacking gel and a separating gel. The stacking gel makes up only approx 10% of the total gel volume, is of a lower percentage acrylamide (usually 2.5%~5%) and a lower pH (6.8), compared with the separating gel, usually containing a higher percentage acrylamide (7%~15%) and at a higher pH (8.8). The lower pH in stacking gel develops a sharp interface between the high mobile Cl^-, and the less mobile $glycine^-$. As the interface moves downward, the protein molecules with mobilities intermediate between Cl^- and $glycine^-$, will be concentrated into a very thin band. When protein molecules reach the junction between a large pore stacking gel and a smaller pore running gel, they are more concentrated.

Sodium dodecyl sulfate (SDS)-PAGE, was introduced in 1967 by Shapiro, et al. and become one of the most popular PAGE of proteins. SDS is an anionic detergent having a 12 carbon hydrophobic chain and the negative charged sulfonic acid group. The proteins interacting with SDS are given rod-like shape, because the detergent disrupts native ionic and hydrophobic interactions. SDS coats proteins with an approx. uniform charge-to-mass ratio (approx. 1.4g SDS/g protein). In addition, disulfide bonds of proteins between Cysteine residues can be cleaved by a reducing agent such as β-mercaptoethanol or 1-, 4-dithiothreitol (DTT). Thus, in denatured and reduced condition, separation of proteins or polypeptides is due solely to differences in size, and this method can be used to determine molecular weight (MW). That is say, a calibration curve can be generated by plotting relative mobilities of standard proteins (i.e., mobility relative to bromophenol blue tracker dye) versus their log MW, and subsequently used to estimate the MW of an unknown protein running on the same gel (Fig. 6-6). The estimation of MW by this method is accurate to approximately 5%~10% of the actual value.

Some proteins having color can be seen directly on a gel, but most proteins in the gel are typically visualized as blue bands by staining with Coomassie brilliant blue (CBB), which is one of the simplest non-radioactive methods. It has a reported sensitivity of 0.1μg protein/band.

Reagents

1. 30% Acrylamide stock solution: 29g acrylamide and 1.0g bisacrylamide dissolve in 100mL dH_2O and through 0.45μm filter. Stable at 4℃ for months. (Caution: it is a neurotoxic.)
2. 1.5mol/L Tris-Cl buffer, pH8.8.
3. 0.5mol/L Tris-Cl buffer, pH6.8.

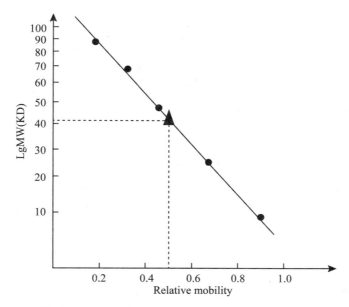

Fig. 6-6 Standard curve for protein MW by SDS-PAGE

4. 10% (W/V) Sodium dodecyl sulfate (SDS).

5. 10% (W/V) Ammonium persulfate (AP) (store at 4℃ for 1~2 weeks).

6. 10% TEMED.

7. Distilled water (dH_2O).

8. 2× Sample loading buffer:

    100mmol/L Tris-Cl, pH6.8;

    100mmol/L DTT;

    20% Glycerol;

    4% SDS;

    0.2% Bromphenol blue (BB).

9. 5× SDS-PAGE running buffer, pH8.3.

    Tris Base: 15.1g;

    Glycine: 72.0g;

    SDS: 5.0g.

    Dissolve in dH_2O up to 1L, storage at room temperature. Dilute to 1× before use.

10. Solutions for Coomassie staining and destaining:

    40% Methanol;

    10% Acetic acid.

    Staining solution contains 0.1% Coomassie Brilliant Blue (CBB) R-250.

11. Protein markers or standards (sometime already prepared in loading buffer):

    Rabbit phosphorylase B(MW 97,400U);

    Bovine serum albumin(MW 66,200U);

    Rabbit actin(MW 43,000U);

    Bovine Carbonic Anhydrase(MW 31,000U);

    Trypsin Inhibitor (soybean)(MW 20,100U);

    Chicken Egg white Lysozyme(MW 14,400U).

Procedure

1. Preparation of the gel:

(1) Gel cassette assembly: clean and completely dry the glass plates, combs, and any other pertinent materials. Assemble the glass plate sandwich of the electrophoresis apparatus according to manufacturer's instruction: place a short plate on top of a spacer plate, insert both plates into the casting frame on a flat surface and clamp.

(2) Pour the separating gel: prepare the separating gel solution for a vertical slab gel with size of 1mm×10cm×8cm as directed in Table 6.2. Add 10% AP and 10% TEMED just prior to pouring gel, stir gently to mix, then immediately apply to the sandwich along an edge of the spacers using a Pasteur pipet until the gel height is lower 1.5cm than the short plate. Allow the gel to polymerize 30~60min by covering gently with a layer of water (1cm).

(3) A sharp optical discontinuity at the overlay/gel interface is visible on polymerization, then pour off the water layer.

(4) Pour the stacking gel: prepare the stacking gel solution as directed in the following table, adding 10% AP and TEMED, and pour. Insert the comb and allow the gel to polymerize completely.

Reagents	12% Separation Gel (mL)	5% Stacking Gel (mL)
Distilled H_2O	4.85	2.77
1.5mol/L Tris-Cl, pH8.8	3.8	—
0.5mol/L Tris-Cl, pH6.8	—	1.25
30% Acrylamide stock solution	6.0	0.83
10% SDS	0.15	0.05
10% AP	0.1	0.05
10% TEMED	0.1	0.05
Total volume	15	5

2. Sample preparation:

(1) Dilute a portion of the protein sample with 2×SDS sample loading buffer in Eppendorf tubes (including preparation of protein markers).

(2) Heat all samples for 3~5min at 100℃ (a boiling water bath) to fully denature the proteins. Briefly centrifuge tube to accumulate protein samples at the bottom of tube.

3. Running the gel:

(1) Carefully remove the comb, and then remove the gel cassette from the casting stand, and place it in the electrode assembly with the short plate on the inside. Place buffer dam plate opposite the gel cassette assembly. Provide a slight pressure on the gel cassette and buffer dam while clamping the frame to secure the electrode assembly.

(2) Place the assembly into the electrophoresis tank and completely fill the inner chamber with 1×running buffer. Check for leaks. Use a gel-loading tip or syringe to pipette buffer into each well to remove bubbles.

(3) Slowly pipette 10~20μL of denatured sample or protein markers at the bottom of the individual wells. (25~50μg of total protein is recommended for a complex mixture.)

(4) Add enough running buffer to the region outside of the frame to cover platinum electrode at the bottom. Cover the tank with the lid aligning the electrodes (black or red) appropriately. Connect the electrophoresis tank to the power supply.

(5) Run the gel at 10mA of constant current until BB dye enters the separating gel, then increase the current to 20~30mA and run until BB dye reaches the bottom of the gel. This can take as long as 1h and depends on the gel composition.

(6) When electrophoresis is complete, turn off the power supply, disassemble the apparatus, a gel knife is used to open the plates for exposing the gel.

4. Proteins detection by staining:

(1) Remove the gel into a container and cover with CBB staining solution. Agitate on a gently orbital shaker for at least 20min. Increase the time to improve detection of possible faint protein bands.

(2) Recycle the CBB staining solution and cover the gel with destaining solution, allow it to agitate slowly, carefully pour off destaining solution and replace with fresh until blue bands and a clear background are obtained.

(3) Destained gels can be rinsed stored in dH_2O or dried by a gel dryer.

Notes

1. The purpose of glycerol in sample loading buffer is to increase the solution density so that the sample will sink beneath the running buffer and stay in the sample loading well, and adding BB (a tracker dye) is to indicate how the electrophoresis is progressing because it migrates ahead of proteins in the gel.

2. Because polymerization of polyacrylamide is inhibited by oxygen, so it must be cast into a sealed mould. Covering with a layer of water can isolate the separating gel from oxygen and form a smoothing gel surface.

3. unpolymerized acrylamide and bis acrylamide are neurotoxic. When preparing and handling gels, use gloves to avoid absorbed through the skin.

Experiment title

Date
Observations and results

(Measure the migration distance of each protein, and construct a calibration curve. Calculate the MW of the unknown protein according to the curve.)

Discussion

1. Acrylamide stock solutions for PAGE are typically made up of _____ and _____, which create pores of different size. Gel Polymerization is catalyzed by free radicals, generated by agents such as _____ in the presence of _____.

2. A discontinuous gel system is used frequently to concentrate samples, it consists of a _____ gel and a _____ gel.

3. In PAGE, the combination of gel pore size and protein _____, _____, and sharp determines the migration of the protein. In SDS-PAGE, separation of proteins is solely on the basic of their _____.

4. All Protein bands after SDS-PAGE can be detected by staining with _____.

5. What was the role of SDS and DTT in the gel loading buffer?

Teacher's remarks
Signature
Date

Chapter 7 Chromatography

In 1903, Russian scientist Mikhail Tswett invented a technique to separate different plant pigments from green leaves. This method was named chromatography (chroma means color and graph means atlas). Chromatography is the laboratory technique that allows separation of different components in the mixture based on their differences of physical and chemical properties including absorption capacity, solubility, molecular shape and size, molecular polarity and binding affinity for a particular ligand or a solid support. Nowadays, chromatography has been developed into a multidisciplinary technology. Due to its high efficiency, high sensitivity and high resolution potency, chromatography is widely applicable in industry, agriculture, biochemistry, chemical engineering, medicine science, and so on.

7.1 Fundamental Principles of Chromatography

7.1.1 General principles of chromatography

Although there are many different types of chromatography, the principle of them for separating of the molecules are almost the same. Generally, there are two phases for a chromatography: stationary phase and mobile phase. In the process of chromatography, stationary phase is kept stationary state, while the mobile phase flows over or through the stationary phase. The phases are chosen so that components of the mixture have different distribution in each phase. Components distributing preferentially in the mobile phase will pass through the chromatographic system faster than those distributing preferentially in the stationary phase. Due to their differences of mobility, components can be differentially eluted from the stationary phase following with the mobile phase, and then be detected and analyzed. This process of eluting is named as "development".

7.1.2 Classification of chromatography

There are many classification schemes for chromatography.

(1) According to the physical nature of the mobile phase, chromatography can be divided into liquid chromatography (LC) and gas chromatography (GC). The stationary phase can also take two forms, solid and liquid, which provides subgroups of LC and GC: liquid-solid chromatography (LSC), liquid-liquid chromatography (LLC), gas-solid chromatography (GSC) and gas-liquid chromatography (GLC).

(2) According to the distribution style, chromatography can be divided into adsorption chromatography, partition chromatography, ion exchange chromatography, gel chromatography and affinity chromatography according to the principle of chromatography.

(3) According to the mode of the stationary phase, chromatography can be divided into column chromatography and plane chromatography according to the modes of separation. Plane chromatography includes paper chromatography, thin layer chromatography, thin film chromatography and so on.

Modern chromatography is often performed in a column format. Here, column chromatography will be discussed in details.

7.1.3 Basic components of column chromatography

Column chromatography consists of a solid stationary phase and a liquid mobile phase. The stationary phase is confined to a column and the mobile phase (a buffer or solvent) is allowed to flow through the solid phase in the column. A basic apparatus for column chromatography systems mainly comprise a buffer reservoir, a chromatographic column, a detector, a recorder and a fraction collector. The buffer reservoir contains liquid mobile phase delivered either by gravity or using a pump. The chromatographic column is the core part of chromatography separation. Generally, it is a glass or plastic column containing stationary phase. The solution flows out from the chromatography column to the automatic fraction collector through a detector, such as the ultraviolet monitor. The signal detected by the detector is recorded by the recorder. The outcome of a chromatography experiment can be shown as a chromatogram. The complete chromatographic system may also contain an autosampler to add samples and a computer to control the system (Fig. 7-1).

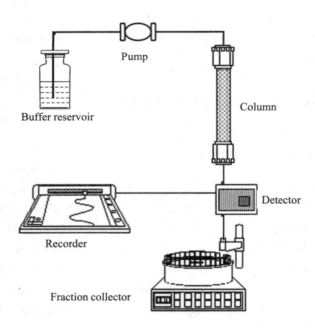

Fig. 7-1 Basic components of column chromatography

7.2 Commonly Used Chromatographic Techniques

7.2.1 Partition chromatography

7.2.1.1 Principle

Partition chromatography is a separation technique based on the different solubility of the components in the mixture between stationary phase and mobile phase.

The different solubility of the solutes between the two phases can be described in terms of the distribution coefficient (K_d).

$$K_d = \frac{\text{Substance concentration in stationary phase}}{\text{Substance concentration in mobile phase}}$$

The molecule with higher K_d retains more in the stationary phase and flows slowly. However, the molecule with lower K_d migrates faster. K_d is dependent on both the temperature and the properties of the solutes and solvents. Under a given temperature, the migration rate of a compound can be described in terms of relative mobility (R_f).

$$R_f = \frac{\text{Distance traveled by sample from origin}}{\text{Distance traveled by solvent front from origin}}$$

R_f is dependent on the K_d of the solutes between the two phases as well as the volume ratio of two phases. Because the volume ratio of two phases is constant under identical experimental condition, the R_f value is only decided by the K_d. Different components in the mixture with different K_d and R_f can be separated after a given time of chromatography. R_f is a constant for a particular compound under the standard conditions. The components can therefore be identified by their characteristic R_f values.

7.2.1.2 Developing reagent

In normal-phase partition chromatography, the stationary phase is polar solvent (such as water bounded to the filter paper in paper chromatography), and the mobile phase used in development is nonpolar solvent (usually combinations of organic and aqueous solvents). In reverse-phase partition chromatography, the stationary phase is nonpolar and the mobile phase is polar.

7.2.1.3 Detection and analysis of components

The colored sample bands or sample spots can be observed during the process of chromatography. But if the sample is colorless, the sample bands or spots have to be colored by spraying reagent for reaction. The R_f value of different components in a sample are calculated, and components can be identified based on their R_f values as well as those of the standards.

7.2.2 Ion exchange chromatography

7.2.2.1 Principle

Ion exchange chromatography is a separation method based on charge differences. There are two types of ion exchange chromatography, namely cation exchange chromatography and anion exchange chromatography. The stationary phases are ion exchangers which are either positively or negatively charged on their surface. The mobile phases are electrolyte solutions with a particular pH and ionic strength. The choice of which type of ion exchanger, as well as the composition of the buffers used in the experiment should be determined in advance.

Take cation exchange chromatography as an example: cation exchange refers to a situation in which the stationary phase carries negative charges and has affinity for cations in solution. When a mixture of charged molecules passes through the column containing the cation exchanger, the molecules with strong positive charges will bind much tightly than those carrying weak positive charges. The bound molecules can be eluted out one by one by an eluting buffer with a suitable pH and ion strength or a pH gradient buffer.

Ion-exchange chromatography can be used to separate mixtures of biological molecules with ionized groups including amino acids, polypeptides, proteins, and nucleotides.

7.2.2.2 Ionic exchanger

The basic requirements of ion exchanger are loose and porous structure or huge surface area, allowing free diffusion and exchange of charged molecules in it, highly insolubility, more exchangeable groups, physical and chemical stability.

Depending on their composition and properties, the ion exchangers can be classified as hydrophobic and hydrophilic ion exchanger. The hydrophobic ion exchangers are made by co-polymerizing styrene with divinylbenzene, and are normally used to separate small molecules. The hydrophilic ion exchangers including cross-linking agarose, cross-linking sephadex, cellulose and polyacrylamide with varying degrees of cross-linking, are normally used to separate biomacromolecules such as nucleic acids and proteins.

Depending on the exchangeable ions and their exchange properties, the ion exchangers can be classified as cation exchanger and anion exchanger. The commonly used cation exchangers are polystyrene sulfonic acid cation exchange resin, carboxymethyl cellulose (CM-cellulose) and CM-Sephadex C-50. The commonly used anion exchangers are diethylaminoethyl cellulose (DEAE-cellulose) and DEAE-Sephadex A-50.

7.2.3 Size exclusion chromatography/ Gel filtration

7.2.3.1 Principle

Size exclusion chromatography, also named as gel filtration, is a separation method based on the differences of molecular weight of components in the mixture. The stationary phases are gel granules that have the effect of molecular sieve (porous matrix). When a solution containing components with different molecular weight passes through the column containing a porous gel matrix, large molecules cannot enter into the pores of gel granules and will be excluded rapidly. Smaller molecules will easily enter the pores of porous gel particles and therefore be eluted out at a slower rate. Fig. 7-2 illustrates the principle of gel filtration.

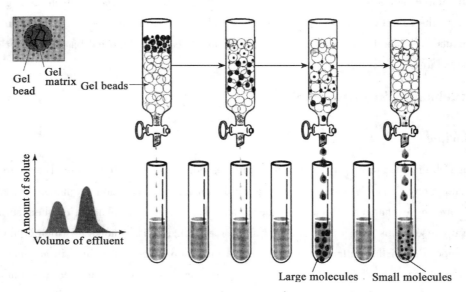

Fig. 7-2 The principle of gel filtration

To assess the extent to which sample molecules can penetrate the pores in the stationary phase, the distribution coefficient (K_d) of a particular solute between the inner and outer solvent are routinely measured. For a given type of gel, K_d is a dependent on molecular size of the solute and is given by:

$$K_d = \frac{V_e - V_o}{V_i} = \frac{V_e - V_o}{V_t - V_o}$$

where V_e (elution volume) is the volume of the mobile phase that is required to elute a particular compound; V_o (void volume) is the volume of mobile phase between the beads of the stationary phase inside the column; V_i (inner volume) is the volume of mobile phase inside the porous beads; V_t is total bed volume, $V_t = V_o + V_i + V_g$ (V_g is the volume occupied by matrix, it is often neglected because it is only a small part of V_t). If the compounds are too large to fit inside any of the pores of gel beads and completely excluded from the solvents within the gel, $K_d = 0$, and they will be eluted at first. If the compounds are small enough to fit inside all the pores of the gel beads, $K_d = 1$, and they will be last eluted. Compounds of intermediate size are partially fit inside some but not all of the pores of gel beads. These compounds will be eluted between the large ("excluded") and small ("totally included") ones, and their K_d values be ranged between 0 and 1.

7.2.3.2 Gel matrix

The gel matrix for gel filtration consists of porous beads with a well-defined range of pore sizes. The commonly used gels include cross-linked dextran polymer (trade mark Sephadex, or Sephacryl), cross-linked agarose (trade mark Sepharose), cross-linked polyacrylamide (trade mark Biogel), and so on.

The dextran gels are porous gel granules cross linked with polysaccharide dextran by chemical method. The gel granules with different degrees of cross linking have different pore sizes. According to the degree of cross-linking, Sephadex may be divided into different types that can be indicated by "water regain", i.e., the amount of water taken up in the completely swollen gel granules per gram of dry gel. For example, if the water regain by one gram of Sephadex is 25 grams, it will be named as Sephadex G-25.

Polyacrylamide gels are prepared by the polymerization of acrylamide and methylene bisacrylamide. By varying the relative proportions of the two monomers, a range of gels with different pore sizes can be obtained. Take Biogel P-6 as an example, it has a molecular weight exclusion limit of 6,000U.

Agarose gels are porous gel granules cross linked by different concentrations of agarose. The commonly used types are Sepharose 2B, 4B and 6B, which represents 2%, 4% and 6% of agarose respectively. This type of gel matrix has satisfactory mechanical property, high resolution and broad separation scope. Like dextran gels, agarose gels cause very little denaturation and absorption of sensitive biological molecules and therefore are widely used in separation of biomacromolecules.

The commercial gel types suitable to differentiate the molecules with the relative range of size are listed in Table 7.1.

Table 7.1   **Commonly used gel matrix for gel filtration**

Gel Matrix	Fractionation Range (U)
Sephadex G-15	50~1,000
Biogel P-2	50~1,000
Sephadex G-25	1,000~5,000
Sephadex G-50	1,500~30,000
Biogel P-10	1,500~30,000
Sephadex G-100	4,000~150,000
Sephadex G-200	5,000~250,000
Sephacryl S300	20,000~1,500,000
Sepharose 4B	60,000~20,000,000

7.2.3.3 Application of gel filtration

Gel filtration can be used to separate mixtures of biomacromolecule, especially enzymes, antibodies and some other globular proteins. It can also be used for desalting, concentration of macromolecular solution, determination of molecular mass of biomacromolecules and removing pyrogenic substances.

7.2.4 Affinity chromatography

7.2.4.1 Principle

Affinity chromatography is a separation method based on the reversible binding of a biomacromolecule to a complementary binding substance (ligand). There is a specific binding capacity between enzyme and substrate analogue, antigen and antibody, avidin and biotin, hormone and its receptors, RNA and its complementary DNA. Affinity chromatography exploits those biospecific relationships to specifically separate the target molecule from a mixture. The biospecific ligand, which is covalently attached to a matrix as the stationary phase, binds to the target molecule within a solution to be analyzed. Ideally, the ligand will bind the target molecule only and all other unbound compounds in the solution will be washed out. Target protein is recovered by changing conditions to favor elution of the bound molecules (typically by varying the pH or increasing the ionic strength or adding inhibitors). The principle of affinity chromatography is shown in Fig. 7-3. Affinity chromatography has been widely used in protein purification, nucleic acid purification, antibody purification from blood serum and separation of a mixture of cells into homogeneous populations.

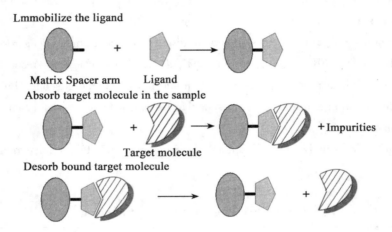

Fig. 7-3 The principle of affinity chromatography

7.2.4.2 Ligand and matrix

A successful separation of affinity chromatography requires a unique biospecific ligand. Therefore, the appropriate selection of ligand is very important. Generally, it is selected according to its magnitude of affinity and specificity to the target molecule. The commonly used affinity systems are listed in Table 7.2.

The specific ligand is immobilized on an insoluble matrix, in a manner which does not interfere with its interaction between ligand and the respective binding substance. This may require the use of a spacer arm, which typically consists of a chain of about 6~10 carbon atoms to facilitate effective binding.

The ideal matrix used in chromatography should have following characteristics:

(1) Insoluble in water but with high hydrophilicity;

(2) No non-specific physical or chemical adsorption;

(3) Enough activated chemical groups, in order to couple with profuse ligand under optimum temperature;

(4) Good physical and chemical stability;

(5) Loose porous network structure, which will enable the macromolecules to pass freely, enabling the effective concentration of ligand.

The most favored and commonly used matrix for affinity chromatography is CNBr-activated Sepharose-4B.

Table 7.2 **Commonly used affinity systems**

Ligand	Respective binding substance
enzyme	substrate analogue, inhibitor, cofactor
antibody	antigen, virus, cell
lectin	cell surface receptor, polysaccharides, glycoprotein
hormone	receptor, hormone carrier protein
biotin	avidin
nucleic acid	complementary base sequence, histone
Ni^{2+}	protein with His 6 tag
Protein A	IgG

7.2.5 High performance liquid chromatography (HPLC)

7.2.5.1 Principle

High performance liquid chromatography (HPLC) has the same principle as other classical chromatography methods. Technically, the mobile phase is transported by high-pressure through the chromatography column, which is padded with particles (3~10 microns) of stationary phase by special method. Its efficiency is much higher than the general liquid chromatography. Simultaneously, a high sensitivity detector is connected to the column, which can examine the flowing liquid continuously. HPLC is divided into analysis mode and preparation mode according to the experimental purpose. The analysis mode is mainly used in the ingredient analysis of the sample, and the preparation mode is mainly used for the preparation of some substances. The stationary phase in HPLC refers to the solid gel matrix and the mobile phase refers to the solvent acting as a carrier for the sample solution. Different stationary phases and mobile phases are selected according to the experimental purpose.

Generally, the smaller the particle size of the stationary phase is, the greater the resolving power of chromatographic column is. However, the particle with smaller size will cause the greater resistance to eluant flow, and will elongate elution time for samples. In order to reduce the elution time, the high pressure pumping systems are needed for the supply of eluant to the column. However, high pressure is not an inevitable pre-requisite for high performance. New smaller particle size stationary phases that can withstand high pressure are available now.

7.2.5.2 Apparatus of a HPLC system and its applications

A HPLC system mainly contains gradient mixers, sampling system with accurate sample valves, high pressure pumps with very constant flow, high efficiency HPLC columns, high sensitive detection system and the computer automatic data processing system.

HPLC has emerged as the most popular, powerful and versatile form of chromatography. HPLC has been developed for many of the types of chromatography including normal-and reverse-phase partition, ion exchange, gel filtration and so on. Preparative HPLC can be used for the separation and purification of proteins, peptides, and polynucleotides. The analytical application of HPLC includes quality control, process control, forensic analysis, environmental monitoring and clinical testing.

Section II Basic Biochemical Techniques

Experiment 6 Separation of Mixed Amino Acids by Cation Exchange Chromatography

Principle

Ion exchange chromatography is a technique to separate mixtures of charged compounds. It relies on the differential electrostatic affinities of charged molecules for a charged stationary phase (ion-exchange resin). Amino acids are ampholytes. The net charge on an amino acid depends on their pI and on the pH of the solution. Different amino acids will carry different charges in a solution with a given pH. The differently charged amino acids can be separated by being eluted from ion exchange columns according to their different binding abilities to the ion exchanger. This experiment uses sulfonic acid cation exchange resin to separate the mixture of acidic amino acid (Aspartic acid, Asp, pI = 2.97) and basic amino acid (Lysine, Lys, pI = 9.74). In the condition of pH=5.3, lysines are positively charged and can bind to cation exchange resin. Aspartic acids are negatively charged and can be washed out without any binding. In the condition of pH = 12, lysines are negatively charged and may then be eluted out. So they can be respectively eluted and separated by changing the pH or the ion strength of eluting solution.

Reagents

1. Resin: sulfonic acid cation exchange resin (732 type).
2. Eluting solution:
(1) 0.45mol/L citric acid buffer (pH5.3): dissolve 285g citric acid and 186g sodium hydroxide in distilled water. Adjust the pH to 5.3 with about 105mL hydrogen chloride. Bring the total volume of the solution to 10L with distilled water (dH_2O).
(2) 0.01mol/L sodium hydroxide (pH12): dissolve 0.4g sodium hydroxide in 1L of dH_2O.
3. Mixture of amino acids: 5mmol/L Asp and Lys in 0.02mol/L hydrogen chloride.
4. Dye reagent (titanium trichloride-ninhydrin solution):
(1) Acetic acid buffer (pH5.5): dissolve 82g sodium acetate in 100mL dH_2O. Then add 25mL acetic acid. Dilute to 250mL with dH_2O.
(2) Dissolve 20g ninhydrin in 750mL ethylene glycol monomethylether, add 1.7mL titanium trichloride, then make the final volume 1L with acetic acid buffer.

Procedure

1. The treatment of resin.

Put about 10g of resin in a 100mL beaker and swell it thoroughly with water. Discard the water. Add 25mL of 2mol/L hydrogen chloride, and stir for 30min. Discard the acid liquor, and wash the resin completely to neutral with dH_2O. Add 25mL of 2mol/L sodium hydroxide to the resin and stir for 30min. Discard the basic liquor, and wash the resin to neutral with dH_2O. Suspense the resin in 100mL citric acid buffer (pH5.3).

2. Packing.

Take a chromatography column with the diameter of 0.8~1.2cm and the length of 10~12cm, put a piece of glass cotton circular cushion at the bottom, and pour the above prepared resin from the top. Close the exit of chromatography column. When the resin sediment appears, let out the excessive solution, add more resin until the height of the resin sediment is about 8~10cm.

3. Equilibrium.

Add pH5.5 citric buffer from the top to wash it until the flowing liquor reaches pH5.5. Close the exit of the column. It is needed about 3~5 times column volume of citric acid buffer. Keep the level of the liquor surface about 1cm higher than that of the resin.

4. Loading and elution.

Open the exit to make the buffer flow out. When the level of the liquor is at almost the same height as that of resin, close the exit. (Do not allow the resin to dry out!) Use a long pipette dropper to add 0.5mL mixture of amino acid to the top of resin directly and carefully, and open the exit to make it flow into the resin slowly. Then wash the column wall twice by adding 0.5mL citric acid buffer. Add citric acid buffer to the top of resin until the level of the liquor is 2~3cm above that of resin. Then elute the resin at the flowing rate of 0.5mL/min. Collect the fractions of 3.0mL in each tube. After collecting 5 tubes, elute the resin with 0.01mol/L sodium hydroxide (pH12) and collect another 7 tubes.

5. Regeneration.

Wash the column with citric acid buffer (pH5.3), and the column can be used again.

6. Amino acid detection.

After collecting the eluent, number the tubes collected. Transfer 0.5mL of eluent from each of the tubes to a clean tube respectively. Add 1mL of pH5.3 citric acid buffer and 0.5mL of dye reagent into the tube. Mix the tubes and incubate for 25min in the boiling water bath. Cool the tubes with running water. Add 3mL of 60% alcohol into the tube and mix. The solution showing royal blue means some amino acid has been eluted out. The degree of the color can show the concentration of amino acid, and it can be determined by measuring the optical density of every tube of eluent (take water as blank and at the wavelength of 570nm). Finally, take the optical density of every tube of eluent as ordinate, and the tube number of eluent as abscissa, then draw an elution curve.

Experiment title

Date
Observations and results

Discussion

1. The stationary phases in ion exchange chromatography are _____, the mobile phases are _____ . Ion exchange chromatography could separate molecules depend on the _____.
2. Why can the mixed amino acids (aspartate and lysine) be eluted from cation exchange resin one by one?

Teacher's remarks
Signature
Date

Experiment 7 Separation of Hemoglobin and Dinitrophenyl (DNP)-glutamate by Gel Filtration Chromatography

Principle

Gel filtration, also named as size exclusion chromatography, separates proteins based on their molecular size. In the experiment, Sephadex G-50 was used to separate the mixture of hemoglobin (Hb, red, MW 64500U) and dinitrophenyl (DNP)-glutamate (yellow, MW 2,000~12,000U). Because of the differences of molecular weight, the two components will be eluted out of the column at different rates. In the process of elution, two colored bands will be observed: the large hemoglobin molecules will be eluted out first as a red band, and the smaller DNP-glutamate molecules will flow out of the column later as a yellow band.

Reagents

1. Sephadex G-50.
2. Mixture of hemoglobin and DNP-glutamate.
3. Distilled water (dH_2O).

Procedure

1. Gel preparation.

Add 4g of Sephadex G-50 to 30mL dH_2O and expand it overnight at room temperature or 2h in boiling. Remove the upper water and small particles.

2. Packing.

Take a chromatography column with the diameter of 1cm and the length of 25cm, put a piece of glass cotton circular cushion at the bottom, and pour about 4cm of buffer into the column. Close the exit of chromatography column. Add the slurry of the prepared gel by gently pouring it down the side of the column. Open the exit when the height of the gel deposited is about 1/3 of the length of the column. Continue stacking until the final surface of the gel bed is 3cm below the top of the column. Close the exit of the column. It is vital that the column be set up properly so that it does not dry out.

3. Loading.

Add the sample to the top of the column bed using a transfer pipette. The tip of the pipette is held against the wall of the column and moved in a circular motion so the sample slowly goes down the gel bed, being sure not to disturb the gel. Open the exit of the column to allow the sample to enter the column. When the level of the liquor is at almost the same height as that of gel, close the exit (do not allow the gel to dry out!). Wash the top of the gel twice 5~10 drops of dH_2O, to ensure all of samples running into the gel.

4. Elution.

Add dH_2O to the column and let it flow. You should be able to see the separation of the colored bands almost immediately. When the first colored band is about to elute from the column, collect the sample with a test tube. Collect the sample with a new tube when the second colored band elutes from the column. Close the outlet and put the end cap back on the bottom of the column when all of the colored bands have eluted from the column.

Experiment title

Date
Observations and results

Discussion

1. We can use gel electrophoresis and gel filtration for separating two biomolecules depend on their significantly different _____.

2. How many different colored bands appear as the sample separates through the column? What color is eluted first? What proteins make up each colored band?

Teacher's remarks
Signature
Date

Section III　Integrative Experiments

Chapter 8   Enzyme Analysis

Enzymes are biological macromolecules with high efficiency and specific catalytic activity in organisms. Almost all metabolic reactions in organisms are catalyzed by enzymes. In the study of enzymes, we should not only understand their structural characteristics and biological functions, but also clarify the factors that affect the enzymatic activity, namely enzymatic reaction kinetics. The enzyme with known function can also be used as analytical tools for qualitative and quantitative analysis of enzyme substrates, coenzymes, activators or inhibitors. Enzymology is of great significance in understanding the law of life activities, clarifying the essence of life phenomena and guiding medical practice and industrial and agricultural production.

8.1   The Activity of Enzyme

8.1.1   Enzyme activity determination

Enzyme activity is a measure of the quantity of active enzyme present and is thus dependent on specified conditions. Under the condition of excess substrate and most suitable other conditions, the initial rate of the enzymatic reaction is proportional to the activity of the enzyme to be tested. Determining the speed of a biochemical reaction catalyzed by the consumption of substrates or the increase of products per unit time. The faster the reaction speed, the higher the enzyme activity.

The changes of substrate or product content in enzymatic reactions can be detected directly or indirectly. Some substrates and products of enzymatic reactions cannot be directly detected, and some auxiliary reagents must be added to achieve the detection purpose. For example, some dye is added to indirectly measure the enzyme activity through color reaction. Usually, the applied substrates are in excess of the demand in a reaction and the amount of reduced substrate only accounts for a little portion of the total substrates. Therefore, it is difficult to determine the consumption of substrates accurately. However, the product newly formed in the reaction can be precisely measured by proper sensitive methods.

The methods selected for enzyme activity assay are dependent on the physical and chemical characters of substrates or products. The commonly used assays for measurement of enzyme activity are chemical assay, spectrophotometric assay, electrochemical assay, fluorimetric assay and radiometric assay.

8.1.1.1   Chemical assay

Chemical method is often a discontinuous assay, in which the enzyme-catalyzed reaction is stopped by adding the enzyme denaturant. The enzyme-catalyzed reaction can also be stopped by heating that can lead to thermal denaturation and inactivity of the enzyme. The method is tedious and time consuming. Moreover, it is not easy to obtain precise result for some rapid reactions.

8.1.1.2   Spectrophotometric assay

Spectrophotometric assay is the most popular method based on the different spectral absorbance between substrate and product at certain wavelength. The change in absorbance can be used as the basis of quantitative analysis in the assay according to the Lambert-Beer's law. If the wavelength is in the visible region, a change in the color of the assay system can actually be seen, and these are called colorimetric assays. Many assays are based on the interconversion of $NAD(P)^+$ and $NAD(P)H$. The reduced form, $NAD(P)H$ has the maximum absorbance at 340nm, but the oxidized form, $NAD(P)^+$ has not this character. Therefore, the reaction catalyzed by the dehydrogenases using $NAD(P)H$ as coenzyme can be measured by monitoring the absorbance change at 340nm.

8.1.1.3 Fluorimetric assay

Fluorometric assay is a method to measure the intensity of fluorescence emitted from substrate, product or coenzyme. For example, NADPH, the reduced form of coenzyme of some dehydrogenases, can emit strong blue-white fluorescence in neutral solution while the oxidized form ($NADP^+$) doesn't emit fluorescence. Many enzymes can be assayed by coupling with an appropriate reaction that involves NADPH. The enzyme activity can be expressed as the change of fluorescence intensity per unit time.

The advantage of fluorometric assay is its high sensitivity which is higher than that of spectrophotometric assay for 2~3 orders of magnitude. Thus, it is extremely suitable for the rapid enzymatic analysis of enzymes or substances with a very low concentration. The disadvantage of fluorometric assays is that the fluorometric reading is not directly proportional to the concentration of substance. Furthermore, the results may vary with the experimental conditions such as temperature, scattering of light, equipment, etc.

8.1.1.4 Radiometric assay

Radiometric assay is a method to measure the isotope incorporated into products or isotope release from substrates. The radioactivity can be measured after separating the substrates from the products, which can represent the increase of products or the decrease of substrates in the reaction. The radioactive isotopes most frequently used in this assays are ^{14}C, ^{32}P, ^{35}S and ^{125}I. Radioactive isotopes assay is extremely sensitive and specific. However, this assay is extremely tedious and dangerous due to its radioactivity, the operator must be very cautious and the experiment should be performed in the special laboratory with the specific permission for isotope usage.

8.1.2 The unit of enzyme activity and specific activity

The enzyme activity unit indicates the relative content of the enzyme, In 1976, the International Enzyme Commission defined that one unit of enzyme activity is the amount of enzyme which will catalyze the transformation of 1mM of the substrate per minute under standard conditions. This unit has a symbol "U". Another unit of enzyme activity is the katal, which is the amount of enzyme that converts 1 mole of substrate per second. So $1U = 1/60$ micro katal $= 16.67$ nano katal (or 1katal $= 6.0 \times 10^7 U$).

The specific activity of an enzyme is another common unit, which is defined as the activity of an enzyme unit per milligram of total protein (expressed in U/mg). Specific activity is equal to the rate of reaction multiplied by the volume of reaction divided by the mass of total protein. Specific activity is usually constant for a pure enzyme. The higher the specific activity of the same enzyme in different products or in different purification processes, the higher its purity is.

8.2 Enzyme Kinetics (Factors Affecting Reaction Velocity)

Enzyme kinetics is the study of the chemical reactions catalysed by enzymes. In enzyme kinetics, the reaction rate is measured and the factors affecting reaction rate are investigated. These factors include substrate concentration, enzyme concentration, temperature, pH, activator and inhibitor, etc. Studying enzyme kinetics in this way can reveal the catalytic mechanism of this enzyme, its role in metabolism, how its activity is controlled, and how a drug or an agonist might inhibit the enzyme.

8.2.1 Effecters of reaction velocity

8.2.1.1 Substrate concentration

The Michaelis-Menten equation illustrates the relationship between initial reaction velocity V_i and substrate concentration $[S]$. As shown in Fig. 8-1, V_i varies hyperbolically with $[S]$. At low $[S]$, V_i is almost directly proportional to $[S]$. With the increase of $[S]$, V_i is no longer proportional to $[S]$. If $[S]$ is continuously increased and is high enough, the active sites of enzyme will be saturated with the substrate and the reaction velocity will reach a maximum, that is, V_i is independent of $[S]$.

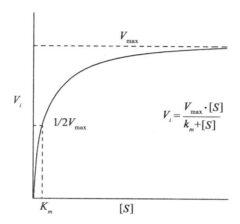

Fig. 8-1 The Michaelis-Menten equation

The V_{max} is the maximum velocity of the reaction. K_m is the Michaelis constant. The K_m of an enzyme is equal to $[S]$ at which the reaction rate at half of V_{max}. So, it is expressed in units of concentration, usually in molar units. K_m is an indicator of the affinity that an enzyme has for a given substrate. When the inhibitor of the enzyme existed, the type of inhibition can be determined by determining V_{max} and K_m for the enzyme.

There are limitations in the quantitative interpretation of a Michaelis-Menten equation plot. A more accurate way to determine V_{max} and K_m is to convert the data into a linear Lineweaver-Burk plot, which is generated by taking the reciprocal of both sides of the Michaelis-Menton equation. V_{max} and K_m are easily determined from the intercept on the Y or X axis (Fig. 8-2).

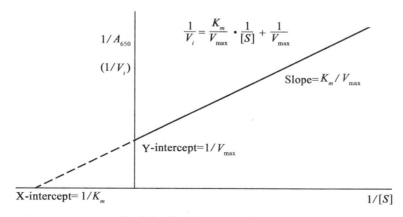

Fig. 8-2 The Lineweaver-Burk plot

8.2.1.2 Enzyme concentration

In order to study the effect of the enzyme concentration on the reaction velocity, the substrate must be present in an excess amount, i.e., the reaction must be independent of the substrate concentration. Under this condition, the enzyme will be saturated with the substrate and the reaction velocity will be directly proportional to the enzyme concentration.

8.2.1.3 Temperature

Temperature displays dual influences on reaction velocity. The reaction velocity approximately doubles for every 10℃ rise in temperature. However, the higher temperature could lead to a thermal denaturation and inactivation of the enzyme. The temperature at which the reaction velocity is the maximum is defined as the optimum temperature of an enzyme-catalyzed reaction, which is normally 35~40℃. However, the optimum temperature of an enzyme is not the characteristic constant and is time-dependent. The optimum temperature decreases as the reaction proceeds. For this reason, enzyme assays are routinely carried out at 30℃ or 37℃, but not always at the optimum temperature.

Although the enzyme activity decreases as the temperature declines, enzymes are generally not destroyed at low temperature. When temperature rises again, the enzyme activity recovers accordingly. To avoid the denaturation of enzyme, enzymes should be kept in refrigerator, and the temperature of the reaction solution should be controlled strictly during measuring of enzyme activity.

8.2.1.4 pH value

Enzymes in general are active only within a limited pH range, and most have a particular pH at which their catalytic activity is maximum, which is defined as the optimum pH. The optimum pH value of most enzymes in animal cells is near neutrality.

The optimum pH is not the characteristic constant of the enzyme and it is affected by factors such as substrate concentration, the type and concentration of reaction buffer and the purity of enzyme. Extremely high or low pH values generally result in complete loss of activity of most enzymes. Thus, in order to maintain a constant enzyme activity, the reaction buffer at the optimum pH should be chosen in enzyme assay.

8.2.1.5 Inhibitor and activator

Any substances that can convert an inactive enzyme into an active one or can increases the enzyme activity in a catalyzed reaction is called an enzyme activator. Enzyme activators can be classified as essential and non-essential activators. Most of the essential activators are metal ions, such as magnesium (Mg^{2+}), potassium (K^+) and manganese (Mn^{2+}), etc. They are essential for enzyme-catalyzed reactions. Any substances that can reduce the enzyme activity without causing denaturation of enzyme are known as enzyme inhibitors. Based on either tightly or loosely bound to the enzymes, inhibitors can be divided into two types, the reversible inhibitors and the irreversible inhibitors.

8.2.2 Initial rate experiments

Optimally, determination of enzyme kinetic parameters is accomplished by measuring production of metabolites, but in cases where authentic standards of metabolites are not available, substrate depletion experiments may provide reasonable estimates. Biochemists usually study enzyme-catalysed reactions using four types of experiments: initial rate experiments, progress curve experiments, transient kinetics experiments, relaxation experiments. Initial rate experiment is by far the most commonly used type of experiment in enzyme kinetics.

For initial rate experiments, when an enzyme is mixed with a large excess of the substrate, the enzyme-substrate intermediate forms rapidly and reaches a steady-state kinetics. The reaction rate is measured in a short time, usually by monitoring the accumulation of the product. Determining the initial rate of the enzyme reaction requires a sufficient substrate concentration (at least 10 times that of K_m) and suitable enzyme concentration. The simplest initial rate experiments involve measuring the time at which a readily detectable substance occurs at the early stages of a reaction. To detect the change of the concentration of one of the components in the reaction, the reaction system should keep the constant of the other conditions—the concentrations of other reactants, the solution volume and the temperature and so on, the time it took for the same reaction was then recorded.

8.3 Enzymatic Analysis

Enzymatic analysis refers to the analytical method using enzymes as tools, which can qualitatively and quantitatively analyze the substrates, coenzymes, activators or inhibitors of enzymes. The enzymatic analysis has high specificity and high sensitivity (10~100mol/L) with simple and convenient operation, it has been widely used to accurately determination of some trace substances in body fluids in clinical chemistry. The commonly used in enzymatic analysis are kinetic analysis, endpoint assay and enzyme-linked immunosorbent assay (ELISA). The enzymatic endpoint method, also called discontinuous determination, is to determine the variable quantity of the substance, products or coenzymes at the end of the transformation reaction which is catalyzed by an enzyme. In the enzymatic analysis of metabolites, spectrophotometry is still the most common detecting method. According to the principle, it is generally divided into single enzyme assay, enzyme coupled assay, enzymatic cycling assay, etc.

8.3.1 Single enzymatic reaction assay

8.3.1.1 Measurement of the substrate amount

Single enzymatic reaction is defined as the reaction catalyzed by one enzyme.

1. Measurement of the substrate consumed

If the substance measured is a substrate in an enzyme-catalyzed reaction, and the substrate can be completely converted to product (when 99% of substrates are converted into products, the reaction is considered to be complete), then the amount of substrate consumed in a reaction can be directly measured by detecting a special absorbance of substrate. For example, the amount of uric acid can be quantitatively measured by uricase. The reaction formula is shown as following:

$$\text{Uric acid} + 2H_2O + O_2 \xrightarrow{\text{Uricase}} \text{Allantion} + CO_2 + H_2O_2$$

Uric acid has two specific absorption peaks at wavelengths 293nm and 297nm, and their molar extinction coefficients are 12.6×10^6 and 11.7×10^6 respectively. Therefore, the amount of uric acid can be directly measured by monitoring the decrease of the absorbance at 293nm and 297nm in the reaction catalyzed by uricase.

2. Measurement of the product formed

If the substance measured is a product and almost all substrates can be converted into products which can be specifically measured, then the amount of substrate can be measured by monitoring the amount of the product produced.

3. Determination of the amount of coenzyme

In a reaction catalyzed by dehydrogenases utilizing NAD^+ or $NADP^+$ as coenzyme, the amount of substrate can be quantitatively measured by monitoring the absorbance change of NADH or NADPH at wavelength 340nm (the molar extinction coefficient is 6.22×10^6). For example, the quantitative measurement of L-glutamic acid is based on measurement of the formation of coenzyme NADH.

$$\text{L-glutamic acid} \xrightarrow[\text{NAD}^+ + \text{H}_2\text{O} \quad \text{NADH} + \text{NH}_4^-]{\text{Glutamic acid dehydrogenase}} \alpha\text{-ketoglutarate}$$

8.3.1.2 Quantitative measurement of coenzymes

Coenzymes play important roles in maintaining the function of enzymes, it is necessary to quantitatively measure the coenzymes. Coenzyme can be measured by single enzymatic reaction. For example, CoA can be determined by using phosphate transacetylase (PTA). The reaction formula is shown below:

$$\text{CoA} + \text{acetyl phosphate} \xrightarrow{\text{Phosphate transacetylase}} \text{acetyl-CoA} + \text{H}_3\text{PO}_4$$

Since acetyl-CoA has an absorbance peak at wavelength 233nm, the formation of acetyl-CoA can be determined by measuring A_{233} and then the amount of CoA can be calculated based on the formation of acetyl-CoA.

8.3.2 Coupled enzymatic reaction assay

When the substrate or the product in single enzymatic reaction can't be easily measured by physical and chemical methods, the quantitative assay can be carried out by coupling another enzyme which acts as an indicator enzyme. For example,

$$A \xrightarrow{E_1} B \xrightarrow{E_2} C$$

In the enzyme-catalyzed reaction above, if A or B is difficult to be measured directly, then the amount of A can be determined by measuring the amount of C using E_2 as an indicator enzyme. The coupled reactions can be divided into two classes.

8.3.2.1 The method using dehydrogenases as indicator enzymes

The most commonly used indicator enzymes are dehydrogenases using NAD^+ or $NADP^+$ as coenzyme. For example,

$$\text{Glucose} \xrightarrow[\text{ATP} \quad \text{ADP}]{\text{Hexokinase}} \text{Glucose-6-phosphate}$$

$$\text{Glucose-6-phosphate} \xrightarrow[\text{NADP}^+ \quad \text{NADPH} + \text{H}^+]{\text{Glucose-6-phosphate dehydrogenase}} \text{6-phosphogluconate}$$

Because the specificity of hexokinase is not very high, the amount of glucose is difficult to be measured directly by single enzyme reaction. Therefore, glucose-6-phosphate, the product of the first reaction, is used as substrate for the second reaction catalyzed by the indicator enzyme, glucose-6-phosphate dehydrogenase. The amount of NADPH product is then easily determined by measuring the absorbance at wavelength 340nm and thus the amount of glucose can be quantitatively measured based on the NADPH produced.

8.3.2.2 The method using other enzymes as indicator enzymes

Except for dehydrogenases, some enzymes can also be used as indicator enzymes in coupled enzymatic reaction. For example, glucose is oxidized by the enzyme glucose oxidase (GOD) to produce gluconic acid and release hydrogen peroxide (H_2O_2). H_2O_2 is further converted to water and oxidized products by coupling with the enzyme peroxidase (POD), which acts as an indicator enzyme.

$$\text{Glucose} + \text{H}_2\text{O} + \text{O}_2 \xrightarrow{\text{Glucose oxidase}} \text{Gluconic acid} + \text{H}_2\text{O}_2$$

$$\underset{\text{(reduced form of pigment)}}{\text{H}_2\text{O}_2 + \text{DH}_2} \xrightarrow{\text{Peroxidase}} \underset{\text{(oxidized form of pigment)}}{2\text{H}_2\text{O} + \text{D}}$$

The oxidized form of pigments such as dianisidine (DAD) and diaminobenzidine (DAB) have absorption peaks at wavelengths range from 420nm to 470nm. Therefore, the amount of glucose can be determined by measuring the formation of DAD/DAB by spectrophotometry.

8.3.3 Enzymatic cycling assay

The enzymatic cycling assay is a method that uses the substrate specificity of the enzyme to continuously amplify the enzymatic reaction products of the target substance. The characteristic of enzymatic cycling assay is repeated reaction of substrate and coenzyme. The increase of the reaction product improves the detection sensitivity. This method only circulates the target substance and reduces the disturbance of coexisting substances, and improves the detection specificity. This method does not need special equipment. Take the determination of bile acid as an example.

$$\text{bile acid} + \text{Thio-NAD} \xrightarrow{3\alpha\text{-HSD}} \text{3-ketosteroid} + \text{Thio-NADH}$$

$$\text{3-ketosteroid} + \text{NADH} \xrightarrow{3\alpha\text{-HSD}} \text{bile acid} + \text{NAD}$$

Experiment 8　Assay the Activity of Alanine Aminotransferase (ALT) in Serum (Mohun's Method)

Principle

The catalytic activities in the plasma enzymes may serve as qualitative or quantitative indexes of tissue damage. Activities of serum alanine aminotransferase (ALT) were determined colorimetrically according to Mohun's method in 1957.

ALT can catalyze the transamination between L-alanine and α-ketoglutarate, the equation is:

$$\text{Alanine} + \alpha\text{-ketoglutarate} \xrightarrow{\text{ALT}} \text{Pyruvate} + \text{Glutamic acid}$$

Then 2,4-dinitro-phenylhydrazine is added for stopping the reaction and marron compounds are formed which response to α-ketoacid. The absorbance of the product formed from pyruvate is bigger than that from α-ketoglutarate at 520nm. Therefore, the activity of the ALT was calculated by measuring the absorbance of the colored product at 520nm, which represent the amount of pyruvate produced in the reaction. In this experiment, one unit of ALT activity is defined as the amount of enzyme needed to produce 2.5μg of pyruvate permL serum after it is incubated with the substrate at 37℃, pH7.4 for 30min.

Reagents

1. Standard pyruvate solution: 200μg/mL, in 0.1mol/L phosphate buffer, pH7.4.
2. Substrate buffer: L-alanine and α-ketoglutarate.
3. 0.1mol/L phosphate buffer (pH7.4).
4. 0.4mol/L NaOH.
5. 2,4-dinitro-phenylhydrazine solution: 0.2g/L in 1mol/L HCl, dissolved by heating, stored in brown bottle, 4℃.
6. Serum sample. (Because hemolysis can cause a mild increase in ALT activity, sample collection and handling is important. Lipidemia also can cause an artificial increase in ALT activity in endpoint or nonkinetic assays. Serum samples can be stored up to 24~48h in a refrigerator without loss of ALT activity. Avoiding of frozen and thawed repeatedly, the ALT activity of serum samples is usually stable.)

Procedure

1. Four test tubes are used and the operation is done according to the following table:

Reagents (mL)	Test (1)	Test Blank (2)	Standard (3)	Standard Blank (4)
Substrate buffer	0.5	—	0.5	0.5
Incubate the tubes in water bath at 37℃ for 5min				
Serum	0.1	0.1	—	—
Pyruvate (200μg/mL)	—	—	0.1	—
Phosphate buffer	—	—	—	0.1
Mix the tubes, and incubate them in water bath at 37℃ for 30min				
2,4-dinitro-phenylhydrazine	0.5	0.5	0.5	0.5
Substrate buffer	—	0.5	—	—

				续表
Mix the tubes, and incubate them in water bath at 37℃ for 20min				
0.4mol/L NaOH	5.0	5.0	5.0	5.0

Mix the tubes, after 20min at room temperature, the absorbance is read within 30min at wavelength 520nm on condition that absorbance is adjusted to zero with distilled water.

Calculation

$$\text{ALT enzyme activity} \atop \text{(Mohun's Unit)} = \frac{A_1 - A_2}{A_3 - A_4} \times \frac{20}{2.5} \times \frac{1}{0.1}$$

Clinical significance

Plasma normally contains a number of specific enzyme molecules. These enzymes can be grouped as functional enzymes such as thrombin, lipoprotein lipase, etc., and nonfunctional enzymes, which include enzymes of exocrine secretions and intracellular enzymes, such as amylase, transaminase and lactate dehydrogenase. The plasma concentration of most enzymes remains constant in a normal individual. It will be altered if there is: ① change of synthesis of enzymes within the cell; ② cellular damage; ③ change in the size of enzyme forming tissue; ④ an alteration in the rate of inactivation and disposal of enzymes; ⑤ an obstruction to a normal pathway of enzyme excretion.

Measurement of serum levels of numerous enzymes is of diagnostic significance. The serum nonfunctional enzyme determinations are particularly helpful in clinical medicine, it may be useful to: ① assess the severity of the organ damage; ② differentiate a particular type of disease; ③ follow the trend of the disease; ④ determine postoperative risk.

The serum transaminase levels are normally low but elevated after extensive tissue destruction. Hepatic cell injury is manifested by elevated serum transaminase activity prior to the appearance of clinical symptoms. Comparable elevations of both AST and ALT are highly characteristic of acute viral, toxic, or non-ethanol drug-induced hepatitis. Hepatocytes contain more ALT than AST, so ALT is a more specific and sensitive indicator of liver damage.

Elevation of the serum transaminase levels may occur in a variety of non-hepatobiliary disorders. However, elevations exceeding 10~20 times the reference are uncommon in the absence of hepatic cell injury. Since the concentration of ALT is significantly less than AST in all cells except hepatocytes, ALT serum elevations are less common in non-hepatic disorders. Following myocardial infarction, AST activity is consistently increased. AST and only occasionally ALT increase in inflammatory skeletal muscle diseases and progressive muscular dystrophy.

Section III Integrative Experiments

Experiment title

Date

Observations and results

Discussion

1. The roles of 2, 4-dinitro-phenylhydrazine in test tube include _____ and _____.
2. Please explain the reason for setting two kind of blank tubes (test blank and standard blank).
3. Please deduce the formula according to the Lambert-Beer's law and the definition of enzyme activity unit.
4. What is the clinical significance for assaying the activity of alanine aminotransferase in serum?

Teacher's remarks
Signature
Date

Experiment 9 Effect of Substrate Concentration on Enzyme Activity- Determining K_m Value for Alkaline Phosphatase

Principle

In this experiment, we measure the value of the K_m of alkaline phosphatase (AKP, ALP). Alkaline phosphatase catalyzes the following reaction:

$$\text{Disodium phenyl phosphate} + H_2O \xrightarrow[\text{pH10.0, 37°C}]{\text{Alkaline phosphatase}} \text{Hydroxybenzene} + Na_2HPO_4$$

The concentration of hydroxybenzene can be obtained through the following reaction:

$$\text{Hydroxybenzene} + \text{Phosphomolybdic acid-phosphotungstic acid in phenol reagent} \xrightarrow[OH^-]{\text{reduction}} \text{Blue compounds}$$

The intensity (Absorbance, A) of the blue compounds correlates directly with the concentration of the substrate (disodium phenyl phosphate) and can be measured at 650nm. The difference of the concentration of disodium phenyl phosphate before and after the reaction equals to the reaction velocity (V). That is, $1/A_{650}$ equals to $1/V$. Plotting $1/A_{650}$ versus $1/[S]$ gives the Lineweaver-Burk plot whose x-intercept is $1/K_m$, whose y-intercept is $1/V_{max}$, and whose slope is K_m/V_{max}.

Reagents

1. Phenol reagent (prepared as the same in Lowry method).
2. 2.5mmol/L disodium phenyl phosphate (substrate solution).
3. Alkaline buffer (pH10.0): Sodium carbonate (Na_2CO_3) 6.36g and sodium bicarbonate 3.36g are dissolved in 1L distilled water (dH_2O).
4. 25mg% alkaline phosphatase solution: 1mg alkaline phosphatase is dissolved in 4mL dH_2O.

Procedure

Six test tubes are used and the operation is done according to the following table:

Reagents (mL)	1	2	3	4	5	blank
2.5mmol/L disodium phenyl phosphate	0.2	0.4	0.6	0.8	1.0	1.0
dH_2O	0.8	0.6	0.4	0.2	—	—
Alkaline buffer	1.0	1.0	1.0	1.0	1.0	1.0
Mix the contents of tubes, incubate at 37°C for 5min						
Alkaline phosphatase	0.1	0.1	0.1	0.1	0.1	—
Mix the contents of tubes, incubate at 37°C for 15min (Accurate!)						
Phenol reagent	1.0	1.0	1.0	1.0	1.0	1.0
Alkaline phosphatase	—	—	—	—	—	0.1
10% Na_2CO_3	3.0	3.0	3.0	3.0	3.0	3.0

Each tube must be mixed immediately after phenol reagent is added. After incubation at 37°C for 15min, A_{650} is detected on condition that blank tube is adjusted to zero.

Calculation

1. To make the graph, take the reciprocal of each of your data points or in other words for each A_{650} you calculate $1/A_{650}$ and for each $[S]$ you calculate $1/[S]$.

Tube	A_{650}	$1/A_{650}$	$[S]$	$1/[S]$
1				
2				
3				
4				
5				

2. Prepare the Lineweaver-Burk plot by plotting $1/A_{650}$ versus $1/[S]$. V_{max} and K_m can be determined from the intercept on the Y or X axis.

Experiment title

Date
Observations and results

Discussion

1. The V_{max} is _____ . The K_m of an enzyme is the substrate concentration at which the reaction occurs at _____ of the maximum rate.

2. K_m is (roughly) an inverse measure of the affinity or strength of binding between the enzyme and its substrate. The lower the K_m, the _____ the affinity.

3. How will the Lineweaver-Burk plotting curve change when competitive inhibitors are added to the reaction system?

Teacher's remarks
Signature
Date

Experiment 10   Enzymatic Endpoint Method-Determination of Total Cholesterol, Triglycerides and Glucose Levels in Serum

Part I   Determination of Total Cholesterol in Serum (CHO-PAP Enzymatic Endpoint Method)

Principle

The cholesterol is determined after enzymatic hydrolysis and oxidation. Cholesterol esters in the serum are hydrolyzed to free cholesterol and fatty acids in a reaction catalyzed by cholesterol esterase. Then the free cholesterol (the produced and already present in the serum) is oxidized by cholesterol oxidase forming hydrogen peroxide. The indicator quinone imine is formed from hydrogen peroxide and 4-aminoantipyrine in presence of phenol and peroxidase. The amount of total cholesterol (TC) in the sample is determined by measurement of the absorbance of the red color at 460~560nm.

$$\text{Cholesterol ester} + H_2O \xrightarrow{\text{Cholesterol esterase}} \text{Cholesterol} + \text{Fatty acid}$$

$$\text{Cholesterol} + O_2 \xrightarrow{\text{Cholesterol oxidase}} \text{Cholestene-3-one} + H_2O_2$$

$$2H_2O_2 + \text{Phenol} + \text{4-aminoantipyrine} \xrightarrow{\text{Peroxidase}} \text{Quinone imine} + H_2O$$

Reagents

1. Sample: serum, heparinized plasma or EDTA plasma.

Sample for lipid profile should be taken after 12h fasting. This is because the chylomicron present in postprandial plasma elevates triglyceride and reduces LDL and HDL cholesterol. The chylomicrons are usually cleared after 6~9h and their presence after 12h is considered abnormal.

2. Standard cholesterol solution: 200mg/dL or 5.17mmol/L.

3. Reaction reagents: pH6.5, 30mmol/L phosphate buffer containing Cholesterol esterase > 150U/L, Cholesterol oxidase > 100U/L, Peroxidase > 5000U/L, 4-aminoantipyrene 0.25mmol/L, Phenol 25mmol/L (Leadman kit).

4. Distilled water (dH_2O).

Procedure

1. Accurately pipette each of the solutions into duplicate (or triplicate) wells in a microplate.

Reagents (μL)	Blank	Test	Standard
Distilled H_2O	5	—	—
Standard solution	—	—	5
Sample	—	5	—
Reaction reagent	200	200	200

2. Mix the wells by pipetting up and down several times. Incubate at 20℃ to 25℃ for 10min or at 37℃ for 5min (the developed color remains unchanged at least 1h). Avoid exposure to direct light. Measure A_{500nm} of the test sample and the standard against the blank using a spectrophotometric microplate reader.

Calculation

$$\text{Concentration of total cholesterol (mg/dL)} = \frac{A_T}{A_S} \cdot 200$$

Clinical significance

Value	Interpretation
< 5.17mmol/L (200mg/dL)	Desirable blood cholesterol
5.17~6.18mmol/L (200~239mg/dL)	Border-line high blood cholesterol
≥ 6.20mmol/L (240mg/dL)	High blood cholesterol

Cholesterol in the blood is present mostly in high density lipoprotein(HDL) and low density lipoprotein(LDL) in form of cholesterol ester and cholesterol.

HDL cholesterol (HDL-C)

HDL-C is measured in the supernatant after the precipitation of apoB-100 containing lipoprotein (LDL) directly from plasma or serum using a polyanion in presence of divalent cation.

LDL cholesterol (LDL-C)

Total cholesterol (TC) = [VLDL cholesterol] + [LDL cholesterol] + [HDL cholesterol]
VLDL cholesterol (VLDL-C) = [plasma triglyceride]/ 5mg/dL
LDL cholesterol = [TC]-[HDL cholesterol]-[triglyceride]/ 5mg/dL

1. Lipid profile.

Lipid profile is a measure of the lipid contents of the blood. It includes measurement of TG, cholesterol (TC, HDL-C, LDL-C, VLDL-C). Cholesterol and TG are insoluble in water and are transported in the blood in lipoproteins, complexes of lipids with specific proteins known as apolipoproteins. There are four major classes of lipoprotein: chylomicrons (CM), VLDL, LDL and HDL. It assumes major significance in diagnosis and management of cases of dyslipoproteinemia.

2. Dyslipoproteinemia.

The causes of dyslipoproteinemia can be genetic or secondary due to a variety of other conditions, including nephrotic syndrome, obesity, alcoholism, hypothyroidism, diabetes mellitus and certain drugs. The clinical complication of such lipid and lipoprotein metabolism disorder can be:

(1) Premature atherosclerosis.

(2) Deposition of lipid into various tissues including dermis (xanthomas).

(3) Elevated triglycerides presents with a risk of pancreatitis.

Hyperlipoproteinemia may be classified into six distinct phenotypes (the WHO classification) according to lipoprotein particles present in excess.

I	Familial lipoprotein lipase deficiency	TG ↑↑ (due to chylomicron presence)
IIa	Familial hypercholesterolemia	LDL-C ↑
IIb	Familial hypercholesterolemia	LDL-C ↑ and VLDL-C ↑
III	Familial type III hyperlipoproteinemia	TC ↑, TG ↑ (presence of ß-VLDL) VLDL-C, TC/ Plasma TG> 0.3
IV	Familial hypertriacylglycerolemia	VLDL-C ↑
V	Familial type V hyperlipoproteinemia	VLDL-C ↑ & presence of Chylomicron

The most important primary hyperlipidemia is familial hypercholesterolemia. The molecular basic of this condition is a functional defection, or a decrease in the number of LDL receptors, which leads to a decrease in clearance of these lipoproteins from the blood, and an increase in cholesterol synthesis.

3. Screening for Coronary Heart Disease (CHD).

Cholesterol level is an important player in atherogenesis and consequently coronary heart disease (CHD). In particular, LDL-C is implicated in atherogenesis and thus termed "bad cholesterol". HDL-C (involved in reverse cholesterol transport from the tissues to liver for excretion) is considered beneficial and rightly named "good cholesterol". The ratio of serum TC to HDL-C has a direct correlation with the risk factor of CHD. The ratios over 5 : 1 in male and 4.5 : 1 in female are considered as high risk of CHD. Therefore in screening for CHD it is necessary to measure HDL-C as well. Since serum lipids value varies from day to day, at least 2~3 measurements should be made days/weeks apart. Experts agree on TC/HDL-C ratios as a better assessment of individual risk of CHD.

Part II Determination of Triglycerides in Serum (GPO Enzymatic Endpoint Method)

Principle

TG in the sample can be hydrolyzed by lipase to release free fatty acids and glycerol. Glycerol is phosphorylated by ATP to glycerol-3-phosphate (G-3-P) in a reaction catalyzed by glycerol kinase (GK). G-3-P is oxidized to dihydroxyacetone phosphate catalyzed by the enzyme glycerol phosphate oxidase (GPO). In this reaction hydrogen peroxide (H_2O_2) is produced in equal molar concentration to the level of TG present in the sample. The indicator is a red dye, quinone imine, formed from hydrogen peroxide, 4-aminoantipyrine and 4-chlorophenol under the catalysis of peroxidase. The concentration of TG in the serum is determined by measurement of the absorbance at 460~560nm.

$$\text{Triglyceride} + H_2O \xrightarrow{\text{Lipase}} \text{Glycerol} + \text{Fatty acid}$$

$$\text{Glycerol} + \text{ATP} \xrightarrow{\text{GK}} \text{Glycerol-3-phosphate} + \text{ADP}$$

$$\text{Glycerol-3-phosphate} + O_2 \xrightarrow{\text{GPO}} \text{Dihydroxyacetone phosphate} + H_2O_2$$

$$2H_2O_2 + \text{4-aminoantipyrine} + \text{P-Chlorophenol} \xrightarrow{\text{Peroxidase}} \text{Red quinone} + 4H_2O$$

Reagents

1. Sample: serum, heparinized or EDTA plasma, obtained after an overnight fasting, the TG in the sample are stable for several days when stored in the refrigerator, storage of the sample at room temperature is not recommended, since phospholipids may hydrolyze and release free cholesterol, which interferes the determination of TG in serum.

2. Standard TG solution: 100mg/dL or 1.14mmol/L.

3. Reaction reagents: pH7.0, 40mmol/L phosphate buffer containing lipases > 150U/L, Glycerol Kinase > 0.4U/mL, GPO > 1.5U/mL, ATP 1mmol/L, Peroxidase > 5000U/L, 4-aminoantipyrene 0.25mmol/L, Chlorophenol 25mmol/L (Leadman kit).

4. Distilled water (dH$_2$O).

Procedure

1. Accurately pipette each of the solutions into duplicate (or triplicate) wells in a microplate.

Reagent (μL)	Blank	Test	Standard
Distilled H$_2$O	5	—	—
Standard solution	—	—	5
Sample	—	5	—
Reaction reagent	200	200	200

2. Mix the wells by pipetting up and down several times. Incubate at 20℃ to 25℃ for 10min or at 37℃ for 5min. Avoid exposure to direct light. Measure A_{500nm} of the test sample and the standard against the reagent blank within 60min using a spectrophotometric plate reader.

Calculation

$$\text{Concentration of total triglyceride (mg/dL)} = \frac{A_T}{A_S} \cdot 100$$

Clinical significance

The following upper limits are recommended for the determination of the risk factor of hypertriglyceridemia:

Suspected form of TG — 1.71mmol/L or 150mg/dL;

Increased form of TG — 2.29mmol/L or 200mg/dL.

TG are transported in the blood as core constituents of all lipoproteins, but the greatest concentration of these molecules is carried in the TG-rich chylomicrons (CM) and VLDL. CM ultimately enter the blood compartment, where most of the TG is rapidly hydrolyzed in the capillary bed by lipoprotein lipase into glycerol and free fatty acids, which are absorbed by adipose tissue for storage or by other tissues requiring a source of energy. A peak concentration of CM-associated TG occurs within 3~6h after ingestion of a fat-rich meal. CM that have been hydrolyzed in the circulation are termed CM remnants. They are taken up by the liver through a receptor-mediated process in which apoE and/or apoB on the CM remnant surface binds to the apoE receptor, LDL receptor or LDL receptor-related protein (LRP). Significant amounts of circulating TG are also transported in VLDL. After synthesis in the liver and secretion into the blood stream, they are hydrolyzed by LRP into VLDL remnants. Then, the majority continues to be hydrolyzed to LDL.

TG measurements are used in the diagnosis and treatment of diseases involving lipid metabolism (hyperlipoproteinemia), various endocrine disorders (diabetes mellitus, nephrosis) and liver obstruction (extrahepatic biliary obstruction). Although the relationship between TG concentration and risk of CHD has not been firmly established, several studies have found increased TG concentrations to be positively correlated with increased risk for CHD. However, whether TG concentration represents an independent risk factor is not clear.

Plasma TG level is determined after an overnight fasting. Elevated TG level may be due to increased levels of VLDL or combination of VLDL and CM. Only when plasma TG concentration >11mmol/L plus eruptive xanthomas, papules appearance and pancreatitis, high TG represents a major risk. Hypertriglyceridemia is commonly associated with obesity, excessive consumption of sugars and saturated fats, inactivity, alcohol consumption, insulin resistance and renal failure.

Part III Determination of Blood Glucose (Glucose Oxidase-Peroxidase Endpoint Method / GOD-PAP Method)

Principle

Glucose is oxidized by glucose oxidase, forming gluconic acid and hydrogen peroxide. The hydrogen peroxide formed is broken down by peroxidase to water and oxygen. The latter oxidizes phenol, which combines with 4-aminoantipyrine to give quinone imine, a red colored complex. The intensity of the red colored complex measured at 500nm (460~560nm) is proportional to the concentration of glucose in the sample.

$$\text{Glucose} + H_2O + O_2 \xrightarrow{\text{Glucose oxidase}} \text{Gluconic acid} + H_2O_2$$

$$H_2O_2 + \text{4-aminoantipyrine} + \text{Phenol} \xrightarrow{\text{Peroxidase}} \text{Quinone imine} + H_2O$$

Reagents

1. Sample: serum, or plasma anti-coagulated by heparin or EDTA.

The serum or plasma should be separated from the whole blood within 1h after collection and stored at 2~8℃ less than 24h.

2. Working reagents: glucose oxidase > 15U/mL, peroxidase > 1.5U/mL, 4-aminoantipyrine 0.25mmol/L, phenol 0.75mmol/L.

The dispensed compound may store 4~8℃ for one month (Leadman kit).

3. Glucose standard solution (5.55mmol/L, or 100mg/dL).

4. Distilled water (dH_2O).

Procedure

1. Accurately pipette each of the solutions into duplicate (or triplicate) wells in a microplate.

Reagents (μL)	Blank	Test	Standard
dH_2O	5	—	—
Standard solution	—	—	5
Sample	—	5	—
Working reagent	200	200	200

2. Mix the wells by pipetting up and down several times. Incubate them in a water bath at 37℃ for 15min. Avoid exposure to direct light. Measure the absorbance of test and standard at 500nm within 30min using a spectrophotometric microplate reader.

Calculation

$$\text{Concentration of blood glucose (mg/dL)} = \frac{A_T}{A_S} \cdot 100$$

If the concentration of blood Glucose > 27mmol/L, please dilute the serum or plasma with the same volume of

0.9% NaCl at first and multiply the result of assaying with times of dilution.

Reference Values

Concentration of glucose in fasting serum/plasma: 70~110mg/dL or 3.89~6.11mmol/L.

Clinical significance

Glucose is a major sugar in blood. Blood glucose level is strictly regulated by hormones. Insulin decreases blood glucose level, whereas other hormones such as epinephrine, growth hormone and glucocorticoid increase glucose level.

Blood glucose levels fluctuate physiologically as well as pathologically. The condition that blood level above the normal reference level is called hyperglycemia and that below the normal level is called hypoglycemia.

Determining the blood glucose level plays very important roles in identification of various glucose metabolism disorders such as diabetes mellitus. Please read "Experiment 2" for more details.

Section III Integrative Experiments

Experiment title

Date
Observations and results

Discussion

    1. The main lipids in the blood are _____ and _____ . They are transported in the blood as lipoproteins, which are complexes of lipids and _____.

    2. Considering the relationship between lipoproteins and atherosclerosis, _____ is implicated in atherogenesis and thus termed "bad lipoprotein". _____ is involved in reverse cholesterol transport and rightly named "good lipoprotein".

    3. What are the main lipoproteins in plasma?

    4. What kind of lipoprotein has the highest triglyceride content?

    5. Why is plasma TG level increased in diabetes mellitus?

    6. Describe briefly the clinical significance of determining blood glucose level.

Teacher's remarks
Signature
Date

Chapter 9　Isolation, Purification and Identification of Protein

Isolation and purification of various types of proteins and peptides are used in almost all branches of biosciences and biotechnologies. Purification of protein is a series of processes intended to isolate one particular protein of interest from some complex biological (cellular) material, and it may be analytical or preparative.

The purposes for which protein separations are carried out differ widely, and accordingly the criteria of purity also differ. For example, the investigation of the structure, composition or amino acid sequence of a protein requires a purified homogeneous sample. The study of protein's structure and function requires their native structures and biological activities are preserved. The preparation of a protein for commercial or biopharmaceutical use may require large quantities of stable, well-characterized products and at a low cost. Most protein separations require that several criteria be met, such as degree of purity, yield and economy of the method. The key to efficient protein purification is to select the most appropriate techniques, optimize their performance and combine them in a logical way to maximize yield and minimize the number of steps required.

This chapter discusses protein separation and characterization techniques and presents a guide to planning and evaluating a protein purification scheme.

9.1　Procedure of Protein Isolation

The purification of proteins is usually necessary for the study of protein structure and function. The isolated target protein should be highly purified and bioactive. One or a few proteins from a complex mixture, usually cells, tissues or whole organisms, can be isolated using a series of combinatorial methods, dependent on the differences of protein's properties, such as size, density, solubility, charge, binding affinity and biological activity.

Protein purification is generally divided into the four stages: ①pretreatment and cell separation; ②cell lysis and cell organelle separation; ③ extraction, separation and purification of proteins; ④ quantification, concentration, drying and preservation. During the purification process, protein integrity must be maintained to prevent protein from losing its biological activity due to acid, alkali, high temperature and violent mechanical action.

9.1.1　The initial preparation step

It is possible to give detailed consideration to the initial preparation steps. Factors to consider at this stage are the choice of source material, extraction and crude fractionation. It is worthwhile to note some general precautions from beginning to the last step in a separation. For example, keeping temperature as low as possible or adding general antibiotics and proteinase inhibitors for retard the growth of micro-organisms and inhibit the activity of proteolytic enzymes; extremes of ionic strength and pH of solution, adding organic solvents and detergents or violent agitation should be avoided because they tend to denature proteins.

9.1.1.1　Choice of materials and pretreatment

The source from which the protein is isolated is the key to the design of a purification process.

Microorganisms, plants and animals can be used as raw materials for the preparation of proteins. Choice of a starting raw material (such as bacteria, animal tissues/cells, and serum/plasma) is mainly determined by the experiment goal. Bacteria as material are normally used to overexpress a specific protein, polypeptide or enzyme for isolation. If selecting an animal active ingredient rich in tissues or cell, the crude materials should be fresh or frozen immediately before treatment, avoiding proteolytic degradation, and minced, degreased and other treatments should be carried out first.

9.1.1.2 Cell disruption and protein extraction

Except the starting material is serum/plasma or culture supernatant, it is necessary to isolate the tissue/cells or sub-cellular fraction and disrupt the cellular material in buffer. Several methods such as repeated freezing-thawing, sonication and homogenization, are useful for disruption of cells. The method of choice depends on how fragile the protein is and how sturdy the cells are.

No matter what kind of methods used broken cells, the proteinase of cells would be released into the solution, causing proteins degradation. To prevent these effects and obtain the best possible protein yield in cell lysis, protease and phosphatase inhibitors are added to the lysis reagents, such as adding diisopropyl fluoride phosphate (DFP), iodoacetic acid, phenylmethylsulfonyl fluoride (PMSF), or selecting a suitable pH, ionic strength or temperature. See Table 9.1.

Table 9.1 **Disruption methods of cell lysis**

Methods	Principles
Low osmotic pressure	Osmotic disruption of cell membrane
Enzyme digestion (e.g., lysozyme)	Cell wall digested, leading to disruption of cell membrane
Chemical solubilization (e.g., SDS)	Cell membrane solubilized chemically
Liquid homogenization	Cell or tissue suspensions are sheared by forcing them through a narrow gap, disrupting cell membrane but not organelles
Mincing/Grinding	Cells dispersed and disrupted in grinding process by shear force using rotating blades
Bead mill	Rapid vibration with glass beads rip cell walls off
French press	Cell forced through small hole at high pressure, shear forces disrupt cells
Ultrasonication	Microscale high frequency sound waves shear cells
Repeated freezing-thawing	Cells ice crystal formation and swelling, broken cell structure

Differential centrifugation is a common procedure used to separate certain organelles from whole cells for further analysis of specific parts of cells. In the process, a tissue sample is firstly lysed to break the cell membranes and mix up the cell contents. The lysate is then subjected to repeated centrifugations, where density separation causes a sediment to form known as "pellet". After each centrifugation the pellet is removed and the centrifugal force is increased. Finally, purification may be done through equilibrium sedimentation, and the desired layer is extracted for further analysis.

The extracts containing the soluble and particulate fractions are then clarified by filtration or centrifugation steps and checked for the distribution of protein. If the desired protein is associated with the particulate fraction, it will be necessary to solubilize it. The proteins in a soluble form are suitable for the next manipulation. Most extracted proteins are dissolved in salt solution and dilute acid or alkali buffer, but some proteins such as lipoproteins are dissolved in organic solvents such as ethanol and acetone.

During the process of protein extraction, there are two main factors affecting the extraction: the solubility of the extracted substance in the solution and the difficulty degree of diffusion from the solid phase to the liquid phase. Decreasing the viscosity of the solvent, stirring and prolonging the extraction time can improve the diffusion speed of proteins and increase the extraction effect. The principle of protein extraction is "a few times", that is, for the same amount of extraction solution, multiple times extraction is much better than one time extraction.

9.1.2 Crude fractionation

The crude preparation after extraction is normally rather dilute, contains a large number of different proteins and has a large volume. The next step chosen normally is concentrating the sample for further processing. Salting out or isoelectric precipitation is very common initial step in protein purification. It is of economy, simplicity and high capacity.

9.1.2.1 Separation methods according to different protein solubility

1. Salting-out of protein

Neutral salt solution has a significant effect on the solubility of protein. Generally, at low salt concentration, with the increase of salt concentration the solubility and stabilization of protein are increased, which is called salting-in. When the salt concentration is increased further, the solubility of proteins are decreased to different degrees according the hydrophobicity of the protein. This phenomenon is called salting-out. Because high concentration of salt ions (such as ammonium sulfate) have strong hydration effect, which can seize the hydration layer of protein molecules and make them "lose water", thus the protein colloidal particles condense and precipitate. When salting-out, the effect is better if the pH of the solution is at the isoelectric point (pI) of protein. Due to the different particle sizes and hydrophilic degrees of various protein molecules, the concentration of salt required for salting out is also different. Therefore, adjusting the neutral salt concentration in the mixed protein solution can precipitate various proteins by stages. Many types of salts have been employed in salting-out. Ammonium sulfate is commonly used as its high solubility and relatively inexpensive.

2. Isoelectric precipitation

The pI of protein is the pH of a solution at which protein has no net charge. Most proteins have a minimum solubility around their pI because an overall charge near zero minimizes repulsive electrostatic forces and cause hydrophobic forces to attract molecules to each other. This is called isoelectric precipitation. The greatest disadvantage to isoelectric precipitation is the irreversible denaturation because it normally occurs at lower pH below physiological values (the pI of most proteins is in the pH range of 4~6), thus this precipitation is most often used to precipitate contaminant proteins, rather than the target protein. However, this method is rarely used alone and can be used in combination with salting-out method.

3. Organic solvent precipitation at low temperature

Addition of a miscible solvent such as ethanol, methanol or acetone to a solution may cause proteins in the solution to precipitate. The principle effect is the reduction of the dielectric constant of water as the concentration of organic solvent increases. With smaller hydration layers, the proteins can aggregate by attractive electrostatic and dipolar forces. Another feature affecting organic solvent precipitation is the size of the molecule. The larger the molecule, the lower percentage of organic solvent required to precipitate. To avoid denaturation of protein in the less polar solvent, the precipitation is usually carried out at subzero temperatures. In operating procedure, the protein solution should be chilled close to 0℃ and the solvent to at least -20℃.

4. Precipitation with organic polymers

Polyethylene glycol (PEG) is a hydrophilic polymer and will precipitate proteins without denaturing them. The behavior of the proteins is rather similar to their behavior in precipitation by organic solvents. PEG of molecular weights 6,000 and 20,000 is most often used for protein isolation. Large proteins are precipitated at lower PEG concentrations than small proteins. After precipitation, PEG is not easy to remove from a protein fraction. As a polymer it will not be dialyzed rapidly. Acetone precipitation of the protein can be used to separate it from the PEG because PEG is acetone soluble. Another practical difficulty in the use of PEG is that it absorbs UV light at 280nm, and also interferes protein determination by the Lowry's assay. However, protein in the presence of PEG can be measured using the biuret assay.

9.1.2.2 Separation methods based on differences in protein molecular sizes

Some common methods such as dialysis, ultrafiltration, ultracentrifugation, and gel filtration are based on differences in protein molecular sizes.

1. Dialysis

Dialysis separates molecules in solution depending on the difference in their rates of diffusion using a semipermeable membrane. Typically, a semipermeable dialysis bag such as a cellulose membrane with pores, is loaded with a solution of several types of molecules and placed in a container of a dialysis solution. Small molecules and water tend to move into or out of the dialysis bag in the direction of decreasing concentration, while proteins have dimensions significantly greater than the pore diameter are retained inside the dialysis bag (Fig. 9-1(a)).

2. Ultrafiltration

Ultrafiltration is similar to dialysis. It forces water and other small molecules to go through the selective permeable membrane using high-pressure by gas pressure or centrifugation. Ultrafiltration is mainly used as a rapid method for concentrating proteins. Membranes with a nominal cutoff size of 3kU to 100kU are most frequently used, a proportion of molecules of sizes close to the cutoff size will pass through, the larger proteins are retained and concentrated in the container. Ultrafiltration can be performed using centrifuge concentrators or stirred-cell apparatus (Fig. 9-1(b)). Dialysis and ultrafiltration are typically used to separate proteins from buffer components for buffer exchange, desalting, or concentration.

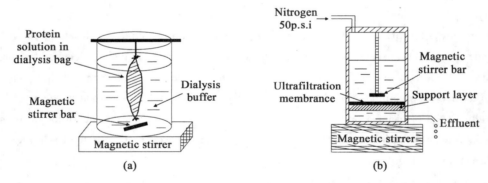

Fig. 9-1 Dialysis (a) and Ultrafiltration (b) of protein solution.

9.1.2.3 Separation methods based on differences in protein mass and density

Ultracentrifugation is valuable for separating biomolecules depending on their differences of mass and density by calculating their sedimentation coefficient (S). The most commonly density gradient ultracentrifugation is used to separate lipoproteins with different S.

9.2 The Selective Purification Steps

At this point the material of interest has been extracted from its source and the major of contaminants has been removed. Unless an affinity step has been used, the methods used so far have had high capacity but relatively poor selectivity. Since the final aim is to obtain the protein of interest in a suitably pure and active form for further study, the next steps need to be more selective.

9.2.1 Separation by electrophoresis

Different types of electrophoresis are able to separate proteins, detect protein purity, estimate MW and pI and obtain information on the physiological activity of the protein. The following electrophoretic methods are commonly used in protein separation.

9.2.1.1 SDS-polyacrylamide gel electrophoresis (SDS-PAGE)

Proteins can be separated largely on the basis of mass by PAGE under denaturing conditions. Sodium dodecyl sulfate (SDS), an anionic detergent, interacts with the proteins to give rod-like complexes. All proteins are masked by negative charge on the SDS and consequently have the same charge/mass ratio. Mercaptoethanol or 1, 4-dithiothreitol (DTT) is also added to reduce disulfide bonds, thus, proteins are in denatured and reduced condition. The mobility of SDS-protein complexes in PAGE is linearly proportional to the log MW (see Chapter 6).

9.2.1.2 Isoelectric focusing (IEF)

Proteins can be separated by isoelectric focusing (IEF) according to their different charges or pI. IEF is a high resolution electrophoretic technique that can be used for preparative work as well as for analytical studies. IEF can readily resolve proteins that differ in pI by as little as 0.01, which means that proteins differing by one net charge can be separated. Each protein will move until it reaches a position in the gel at which the pH is equal to the pI of the protein.

9.2.2 Separation by chromatography

9.2.2.1 Ion-exchange chromatography

Ion-exchange chromatography (IEC) is one of the most popular selective purification methods. IEC can separate proteins based on the reversible ionic interaction between charged proteins and an oppositely charged ion exchanger. It is a very high resolution separation with high sample loading capacity. Proteins bind to charged matrix as they are loaded onto a column, then the elution is usually performed by stepwise or continuous gradient increases in salt concentration (ionic strength) or adjusts the pH of the solvent. Target proteins are collected in a purified, concentrated form. DEAE-Sephacel, DEAE-Sepharose CL-6B are most common ion exchangers, which have high capacity for proteins, excellent resolution and flow rate, and regeneration can be performed.

9.2.2.2 Size exclusion chromatography/ Gel filtration

Size exclusion chromatography (SEC), also called gel filtration, separates molecules on the basis of their size and shape. It is important to choose the different-sized porous beads with the correct selectivity. Sephadex, Sepharose and Bio-gel are commonly used commercial beads. When the protein mixture is added to the beads on the column, large molecules flow more rapidly through this column and emerge first, molecules that of a size to occasionally enter the beads and flow out at an intermediate position, and small molecules, which take a longer, tortuous path, will exit last (Fig. 9-2). Usually proteins are detected as they are coming off the column by their absorbance at 280nm.

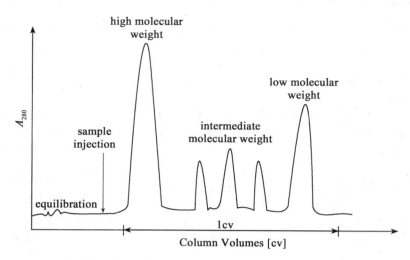

Fig. 9-2 Typical gel filtration elution in protein separation.

9.2.2.3 Affinity chromatography

Affinity chromatography (AC) is the most powerful of the chromatographic techniques. The separation of protein is based on a reversible interaction between a protein and a specific ligand attached to a chromatographic matrix (such as antigen and antibody, enzyme and substrate, or receptor and ligand). After loading the sample onto an affinity column, proteins specifically bind to a complementary binding substance. Unbound material is washed away, then desorption of the target protein is performed specifically, using a competitive ligand, or non-specifically, by changing the pH, ionic strength or polarity. Protein is collected in a purified, concentrated form.

9.2.2.4 Hydrophobic interaction chromatography

Hydrophobic interaction chromatography (HIC) separates proteins with differences in hydrophobicity. The separation is based on the reversible interaction between a protein and the hydrophobic surface of a matrix. A certain concentration of ammonium sulfate just below used for salting out is the most likely to promote binding of proteins to a hydrophobic gel. Fractionation of bound proteins is frequently achieved by decreasing the ionic strength or decreasing the polarity of the eluent (e.g., with ethylene glycol). Phenyl Sepharose CL-4B is an electrically neutral gel for HIC with hydrophobic groups attached by stable ether linkages.

9.2.2.5 High performance liquid chromatography

Using of a technique called high performance liquid chromatography (HPLC) can improve substantially the resolving power of all of the column chromatography with the use of conventional gravity flow column. In HPLC, the media is made up of finer material (e.g., non-compressible resins) and more strong columns is produced with metal, hence, pressures of 5,000~10,000psi are applied to force solutions rapidly through the column to obtain adequate flow rates. Because of high resolution as well as rapid separation of HPLC, it is the most powerful and versatile form of chromatography.

Up till now, there is no single or a set of ready-made methods to extract any protein from complex mixed proteins. Therefore, several methods are often used in combination.

9.3 Crystallization, Concentration, Drying and Storage of Protein

9.3.1 Crystallization

Crystallization, the process of solidifying from solution, is often the final stage in purifying and studying proteins, particularly enzymes. If a solution saturated at some temperature is cooled, the dissolved component begins to separate from the solution. Because the solubilities of two solid compounds in a particular solvent generally differ, it often is possible to find conditions that only one of the components of a mixture crystallizes alone, while the more soluble components remain dissolved.

In order to obtain crystallized protein with a high purity, the crystallization should take place slowly. If solidification is rapid, impurities can be entrapped in the solid matrix. Sometimes, it is necessary to add a seed crystal to the solution to provide a solid surface for beginning the crystallization process.

Except as a method of purification, crystallization is also useful for confirmation of protein homogeneity, determining tertiary structure of protein by X-ray diffraction techniques and storage of purified protein.

9.3.2 Concentration

Concentrative dialysis can be used to remove solvent molecules so that samples can be condensed. The solution is placed into dialysis bag, surrounded by absorbers, e.g., sucrose, PEG or Bio-Gel concentrator resin. Ultrafiltration by ultracent and minicent ultrafiltration cartridges provides an easy method for protein concentration.

Protein precipitation is widely used in downstream processing of biological products in order to concentrate proteins and purify them from various contaminants. The underlying mechanism of precipitation is to alter the solvation potential of the solvent, more specifically, by lowering the solubility of the solute by addition of a reagent. Commonly used reagents include trichloroacetic acid, acetone, ethanol, ammonium sulfate, polyethylene glycol, etc. The first reagent often leads to protein denaturation, while the latter reagents can still maintain protein activity.

9.3.3 Vacuum drying

Lowering the vapor pressure at the given temperature lowers the boiling points of liquids, including that of the solvent. This allows the solvent to be removed without excessive heating. This technique is also called vacuum distillation and it is commonly carried out in the form of the rotary evaporator.

Normally, the methods used for protein drying have vacuum drying and freeze drying (also known as lyophilization). Vacuum drying is the final result of vacuum evaporation. Freeze drying can also be used as a late-stage purification step, because it can effectively remove solvents. Furthermore, it is capable of concentrating substances with low molecular weights that are too small to be removed by a filtration membrane.

9.3.4 Storage

Protein can be preserved in dry powder and liquid form. Both dry powder and liquid protein should avoid long-term exposure to air to prevent microbial contamination. Temperature has a great influence on the stability and biological activity of proteins. Therefore, proteins are usually stored at low temperature ($0 \sim 4°C$), and the storage time should not be too long.

Normally recommendations for storage conditions of purified proteins include: ①store in high concentration of ammonium sulfate (e.g., 4mol/L); ②freeze in 25%~50% glycerol, especially suitable for enzymes; ③sterile filter to avoid bacterial growth; ④lyophilization. The extent of storage "shelf life" can vary from a few days to more than a year and is dependent on the protein nature and the storage conditions. For example, proteins stored in solution at 4℃ can be dispensed conveniently but require more diligence to prevent microbial or proteolytic degradation; such proteins may not be stable for more than a few days or weeks. By contrast, lyophilization allows for long-term storage of protein (several years), but the protein must be reconstituted before use and may be damaged by the lyophilization process.

9.4 Identification and Analysis of Protein

For determining the effect of a protein purification scheme, one requirement is to measure the purity of protein, another is to evaluate yield and specific activity of protein at each purification step. In addition, the assessment of purity should be considered within the overall economy of the purification scheme, therefore, the methods should be quick and efficient.

9.4.1 Identification and purity determination of protein

If the protein has a distinguishing spectroscopic feature, it can be used to detect and quantify the specific protein. If antibodies against the protein are available, then Western blotting and ELISA can specifically detect and quantify the amount of desired protein. Some proteins function as receptors, can be detected during purification steps by a ligand binding assay. Throughout the purification scheme it is important to check the purity of the samples. The most general method to monitor the purification process is by running a SDS-PAGE in the different steps.

9.4.2 Purification yield

In order to evaluate the process of multistep purification, the amount of the specific proteins has to be compared to the amount of total protein. Some of the available methods for measuring protein concentration are described previously, such as Biuret reaction, Lowry assay, Bradford assay, BCA method or UV absorption (A_{280}) (see Chapter 5 Experiment 1). However some reagents used during the purification process may interfere with the quantification. For example, imidazole (commonly used for purification of His-tagged recombinant proteins) is an amino acid analogue and will interfere with the BCA assay at low concentration.

9.4.3 Activity of purified protein

During a protein purification procedure, the most important thing to follow is the recovery of the protein by enzyme activity or bioactivity. If the protein is an enzyme, it required a method for determining the activity of enzyme. The use of zymogram techniques enables detailed information of enzyme activity. For detecting the immunological activity of the protein, immunoelectrophorosis is a good technique that is extremely useful for quantitative measurement, and also for immunological status.

9.4.4 Analysis of purified protein

Purification of protein is the basis of further study of structure and function, or as commercial or biopharmaceutical use. As mentioned above, some specific methods are directly applied to the large-scale separation and identification for a large number of proteins, which are the key experimental techniques in proteomics: 2-dementional electrophoresis (2-DE) combined with mass spectrometry (MS), which allows rapid high-throughput identification of proteins and sequencing of peptides; Protein microarrays, which allow the detection of the relative levels of a large number of proteins present in a cell, and two-hybrid screening, which allows the systematic exploration of protein-protein interactions (see Chapter 3).

Experiment 11 Isolation and Identification of Serum IgG

Immunoglobulin G (IgG) is the most abundant immunoglobulin and is approximately equally distributed in blood and in tissue liquids, constituting more than 70% of serum immunoglobulins in humans. Numerous approaches are well known for immunoglobulin isolation and purification. Here, a scheme for isolation and purification of IgG is described. Ammonium sulfate fractionation is the first procedure, followed by gel filtration for desalting, and DEAE-ion exchange chromatography is applied for further purification of IgG. Then SDS-PAGE and immunoelectrophoresis are used for identification of IgG in this experiment.

Part I Isolation and purification of IgG

Principle

Salting-out is a process that protein precipitates at high salt concentrations. It can be used to fractionate proteins because the salt concentration of protein precipitation differs from one protein to another. Ammonium sulfate fractionation normally is the first purification step, and it can isolate nearly all of crude immunoglobulins from serum. The semi-saturated ammonium sulfate can precipitate globulins, while albumin is soluble. Then 33% saturated ammonium sulfate is used to precipitate out IgG from globulins. After several times of salting out, yielding about 40% pure preparation, enriched IgG is also very stable and suitable for long-term storage.

For further purification, IgG which mixed with ammonium sulfate should be desalted by dialysis or gel filtration. Dialysis is somewhat time-consuming. Gel filtration is a simple, gentle method for separating salt molecules and large proteins on basis of their size. Sephadex G-25 or G-50, as a kind of chromatographic matrix of porous beads, has been widely used. Smaller molecules enter into the pores of the beads and move through the column more slowly, while proteins enter less or not at all and thus move through more quickly.

Coupled with ion exchange chromatography (IEC), IgG will be purified from other contaminating proteins based on differences between the surface charges of the proteins. The ionic interaction occurs between the charge protein and an oppositely ion exchangers. IgG has a higher or more basic pI than most serum proteins, it is positively charged in the pH7.4 PBS buffer, while DEAE-cellulose is a kind of anion exchangers, therefore, IgG does not bind to column and directly elute from the IEC column. Finally, IgG maybe up the level of purification to more than 100-fold, and the yield normally is about 50% in this scheme.

Reagents

1. Saturated ammonium sulfate solution.

Dissolve 800~950g $(NH_4)_2SO_4$ in 1L of hot distilled water (dH_2O), cool to room temperature and adjust pH to 7.2, store at 4℃.

2. 0.9% NaCl.

3. Sephadex G-25 or G-50.

4. 0.5mol/L NaOH.

5. 0.01mol/L phosphate buffer saline (PBS, pH7.4).

6. 0.01mol/L phosphate buffer saline containing 2mol/L NaCl, pH7.4.

7. DEAE-cellulose.

8. 0.5mol/L HCl.

9. 1% $BaCl_2$.

Procedure

1. Salting-out.

(1) Put 10mL serum and 10mL 0.9% NaCl into a 100mL beaker, and mix thoroughly.

(2) Add saturated ammonium sulfate solution 20mL dropwise to produce 50% saturation. Keep stirring on the magnetic stirrer for 30min to precipitate immunoglobulin completely.

(3) Centrifuge at 3000r/min×20min. Discard the supernatant (mainly contains albumin). The precipitate contains all kinds of globulins.

(4) Dissolve the precipitate in 10mL of 0.9% NaCl and transfer to a 50mL beaker.

(5) Add saturated ammonium sulfate solution 5mL dropwise to produce 33% saturation. Go on stirring 30min for precipitation of IgG.

(6) Centrifuge at 3000r/min×20min. Discard the supernatant (mainly contains α-and β-globulin). IgG mainly exists in the precipitate.

(7) Repeat the above steps 3 times to remove coprecipitation (it will be contaminated with other proteins).

(8) Wash the precipitate with 33% saturated ammonium sulfate solution 3 times.

2. Desalting IgG — Gel Filtration.

(1) Preparation of gel: place about 15g Sephadex G-25 or G-50 into a 1L beaker, add 500mL dH_2O for 24h, or boil gently for 2h. After cooling, discard the supernatant and floating small particles, repeat washing 3~4 times.

(2) Packing column: pour the above treated gel into a 1.0cm×50cm glass column. The gel mixture will deposit and distribute naturally.

(3) Equilibration of column: attach the constant flow pump and adaptor, and make the flow rate of dH_2O be 2mL/min at least. After 30min, the column is equilibrated and ready for use.

(4) Connecting the fraction collectors.

(5) Loading sample: completely dissolve precipitate with 1mL 0.9% NaCl, then load sample onto the column using a dropper, when the sample fluid just enter into gel surface, add dH_2O onto the column. Note: Do not disturb the surface.

(6) Collection and detection: elute protein using dH_2O about 2mL/min flow rate, collect 2mL/tube fractions throughout this step. The automatic detector will detect while IgG flow out gradually and then mix them together. Later, $BaCl_2$ solution is used to check the elution of ammonium sulfate.

(7) Regeneration of column: elute the column with dH_2O. When the elution peak returned to baseline and no ammonium sulfate in elution buffer detected by $BaCl_2$, the column can be used again.

3. DEAE-cellulose Ion Exchange Chromatography.

(1) Preparation and packing column: soak 15g DEAE-cellulose with 250mL 0.5mol/L HCl for 40—50min, wash it to pH4 with dH_2O. Then rinse in 250mL 0.5mol/L NaOH for 50min and wash to pH8.0 with dH_2O. Finally equilibrate the column with 0.01mol/L PBS (pH7.4) and pack it into a 1.0cm×50cm glass column. Attach the constant flow pump and adaptor.

(2) Loading and collection: load 1mL desalted IgG solution, elute the column with 0.01mol/L PBS using 1mL/min flow rate. Collect 4mL eluent in each fraction tube. The first protein peak is purified IgG.

(3) Regeneration of DEAE-cellulose column: pass through 2~3 column volumes PBS containing 2mol/L NaCl, and wash thoroughly in PBS until A_{280} of the effluent is less than 0.01, then add preservative and store at 4℃ for the next use.

Part II Identification of IgG product

Principle

After using a purification procedure, it is necessary to check the purity of IgG product and identify its specificity of the immune response. One of simplest methods for assessing purity of IgG fraction is by SDS-PAGE under non-reducing or reducing conditions. If proteins are running in SDS-PAGE without DTT pretreatment, the gel electrophoresis will separate native proteins according to difference in their charge density, and pure IgG will show one band on the gel after staining. If samples are electrophoresed under reducing conditions (adding DTT), staining IgG consists predominantly of two bands comprising heavy (~50kU) and light (~22kU) chains without other contaminating proteins.

Immunoelectrophoresis (IE) that combines electrophoresis and immunodiffusion, is a binary method of determining the levels of three major immunoglobulins: IgM, IgG, IgA. The sera proteins in a well are first separated by horizontal agarose gel electrophoresis on the basis of their different charge-to-mass ratios. IgG products are introduced into narrow troughs parallel to the separated sera antigens. Diffusion of both antigen and antibody takes place, and at a specific ration of antigen and antibody, a precipitin arc of antigen-antibody binding is formed on the plate, indicating the presence of specific IgG. The size, location, and shape of the precipitin arc, as compared to the control, are indications of the amount of protein in the test sample.

Reagents

1. Acrylamide stock solution: 29g acrylamide and 1.0g bisacrylamide dissolve in 100mL dH_2O. Stable at 4℃ for months. Caution: it is a neurotoxin.

2. Separating gel buffer: 1.5mol/L Tris-Cl, pH8.8.

3. Stacking gel buffer: 0.5mol/L Tris-Cl, pH6.8.

4. 10% (w/v) Sodium dodecyl sulfate (SDS).

5. 10% ammonium persulfate (AP) (store at 4℃ 1~2 weeks).

6. 10% N, N, N', N'-tetramethylethylenediamine (TEMED).

7. Distilled water (dH_2O).

8. 2× sample loading buffer: Contain 100mmol/L Tris-Cl, pH6.8; 20% glycerol; 4% SDS; 0.02% Bromphenol blue (BB); 100mmol/L DTT (nonreducing buffer without DTT).

9. Electrophoresis buffer: weigh out 3.0g of Tris base, 14.4g of glycine and 1g SDS to make 1L. The final pH should be 8.3.

10. Solutions for CBB staining and destaining:

Destaining solutions: 40% methanol and 10% acetic acid;

Staining solution contains 0.1% Coomassie Brilliant Blue R-250 in destaining solutions.

11. Protein markers.

12. Purified IgG or column fraction samples; serum as a control.

13. 1% agarose gel.

14. 5×Tris-ethylbarbital electrophoresis buffer: 2.24g ethylbarbital acid, 4.43g Tris, 0.053g Ca^{2+} saline of lactate, 0.065g NaN_3, in 100mL dH_2O.

Procedure

1. SDS-Polyacrylamide Gel Electrophoresis (PAGE) (see Experiment 5 in detail).

(1) Preparing a vertical slab polyacrylamide gel (7.5% separation gel and 3% stacking gel) as described in the following table:

Reagents	3% Stacking Gel (mL)	7.5% Separation Gel (mL)
Acrylamide stock solution	1.0	2.5
Separating gel buffer	—	2.5
Stacking gel buffer	1.25	—
dH$_2$O	2.6	4.75
10% SDS	0.05	0.1
10% AP	0.05	0.1
10% TEMED	0.05	0.05
Total volume	5	10

(2) Preparing and loading samples: determine IgG concentration by Bradford method or calculate by A_{280} (a 1mg/mL IgG solution will have an A_{280} of 13.6). Dilute a portion of IgG sample with 2×SDS sample loading buffer and heat for 3~5min at 100℃. Loading 10~15μg IgG per well.

(3) Running the gel: assemble gel cell and connect the electrodes to a power pack, turn on power supply to run the gel using 10mA of constant current at start, 30mA at end for a 1.0mm gel until BB dye reaches the bottom of the gel. Turn off the power.

(4) Proteins detection by staining: remove gel from plates carefully, and stain with CBB staining solution for 1h (gently rocking) or preferably overnight. Then the gel is destained with destaining solution until blue bands and a clear background are obtained.

2. Immunoelectrophoresis (IE) of IgG product.

(1) Preparation of 1% agarose gel with electrophoresis buffer. Spread gel in horizontal glass plate and let the gel to cool, puncture wells with 4mm diameter drilling equipment and make the trough with razor according the Fig. 9-3.

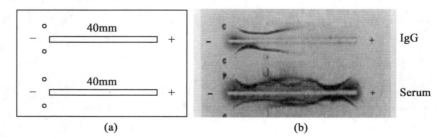

Fig. 9-3 Preparation of agarose gel plate and result of immunoelectrophoresis

(2) Running electrophoresis: add 4%~5% serum, which is diluted with electrophoresis buffer containing 0.01% bromophenol blue (BB) into the well and put the gel plate in the electrophoresis chamber. Communicate two ends of plate to the buffer with wetted filter paper and put the cover on the chamber. Turn on power supply, run the gel at the 6V/cm for 1h with a migration distance of 35mm (when BB dye reaches the end of the gel).

(3) Diffusion: remove the agar plate from the chamber and put it on a flat surface. Remove the buffer in trough and add appropriate amount of IgG (0.1~0.2mL) into the trough. Incubate the gel plate at 30℃ in a humidity chamber containing a moist paper wick for at least 12h. Transfer the plate to 0.85% saline, which can stop the diffusion and wash out unbound proteins, then a precipitin arc of antigen-IgG binding is visible on the plate by naked eyes or staining with CBB dye after drying the gel.

Section III Integrative Experiments

Experiment title

Date
Observations and results

(Calculate the yield of IgG product after IEC. Check the purity of IgG by SDS-PAGE and observe the precipitation arc of IgG product which is formed in immunoelectrophoresis.)

Discussion

1. Describe briefly the main methods for separation and purification of proteins.
2. Describe the methods for identification of purified protein.
3. The most general method to monitor the purification process is by running a _____ in the different steps.

Teacher's remarks
Signature
Date

Experiment 12 Purification and Identification of Glutathione S-Transferase Fusion Protein

Fusion proteins are commonly used for producing foreign proteins in *Escherichia coli* (*E. coli*). The glutathione S-transferase (GST) fusion system using pGEX expression vector has proven successful in producing properly folded and biologically active proteins or polypeptides. Each pGEX vector contains an open reading frame encoding GST, followed by unique restriction endonuclease site (*Bam*H I, *Sma* I, *Eco*R I) and termination codons. If a target gene is inserted into pGEX-2T or pGEX-3X vector at the end of the GST gene, and transferred to *E. coli*, GST fusion protein will be high-level expressed on induction with isopropy-β-D-thiogalactoside (IPTG). Using some convenient methods for specific purification of GST fusion protein from cellular contaminants, and the site-specific protease (such as thrombin) enzymatic cleavage of GST protein, the target proteins are readily amenable to the study of their biological activities and/or interactions.

Part I Expression and Purification of GST Fusion Protein

Principle

GST regions often can be exploited in purifying the target protein. With the use of glutathione as a specific ligand for the GST protein, affinity precipitation is a simple and the most powerful technology for purification of GST fusion protein. Because glutathione is covalently coupled on a solid support (sepharose, agarose or certain water-soluble polymers), combining the affinity interaction and the precipitation, GST-glutathione complex can precipitate and separate from other components in solution. Then, GST fusion protein can elute by washing with elution buffer containing excess glutathione, or the target proteins are obtained by enzymatic or chemical cleavage from GST fusion protein.

Reagents

1. pGEX-2T vector.
2. Competent cells of *E. coli*.
3. LB plates containing 50μg/mL ampicillin (Amp).
4. LB medium containing 50μg/mL ampicillin (Amp).
5. 1mol/L isopropy-β-D-thiogalactoside (IPTG): weigh 2.38g IPTG and add ddH$_2$O to 10mL, then filter with 0.2μm of filter membrane, aliquot and store at -20℃.
6. 0.01mol/L sodium phosphate buffer (PBS, pH7.4).
7. Suspension solution of glutathione agarose beads.
8. 10% triton X-100.
9. GST wash buffer: 50mmol/L Tris-Cl (pH7.5) /150mmol/L NaCl.
10. GST elution buffer: 50mmol/L Tris-Cl (pH8.0)/5mmol/L reduced glutathione (GSH) (fresh prepare, pH7.5).
11. Reagents for SDS-PAGE (see Experiment 13 Part II).

Procedure

1. Subclone the chosen DNA fragment into pGEX vector, transform into competent *E. coli* and screen the colony containing transformant on LB/Amp plates, pick it up and inoculate it in 2mL of LB/Amp medium. Include

a control for setting the transformed bacteria with pGEX vector. Grow cultures with vigorous agitation in a shaking incubator at 37℃ until visibly turbid (3~5h).

2. Induce fusion protein expression by adding 1mol/L IPTG to a final concentration of 0.1mmol/L. Continue shaking at 37℃ for another 3~6h.

3. Transfer the liquid culture to a microcentrifuge tube. Centrifuge the culture 10,000g×15s, and discard supernatant. Resuspend pellets in 300μL ice-cold PBS. Transfer 10μL to another centrifuge tube for checking the expression of GST/fusion protein by SDS-PAGE.

4. Lyse cells by using a sonicator with a 2mm probe and then centrifuge at 10,000g×5min at 4℃. Transfer the supernatant to a new centrifuge tube.

5. Add 50% 50μL glutathione agarose beads suspension to supernatant and mix gently > 2min at room temperature. Add 1mL PBS and spin at 10,000g×5s. Collect beads and discard supernatant. Wash the beads 2 times with PBS.

6. Add an equal volume of 1×SDS sample loading buffer to above beads, and 30μL to the 10μL resuspension cells (from step 3), heat for 3min at 100℃, centrifuge briefly. After loading, run the 10% SDS-PAGE. GST protein from pGEX and fusion protein can be visualized by staining with Coomassie brilliant blue.

7. Inoculate a colony of the pGEX transformant into 50mL LB/Amp medium, grow 12~15h at 37℃ in a shaking incubator.

8. Dilute above culture 1 : 10 into 500mL fresh LB/Amp medium, culture at 37℃ for 1h, add 1mol/L IPTG to 0.1mmol/L (final) and keep on cultivating for 3~7h.

9. Centrifuge cells for 10min at 5,000g, discard supernatant and resuspend the pellet with 10~20mL ice-cold PBS.

10. Lyse cells by sonicator with 5mm diameter probe within 30s, adjust frequency and intensity of sonication so lysis occurs in 30s, without frothing.

11. Add 10% Triton X-100 to final concentratine with 1% and mix. Centrifuge cells at 10,000g×5min at 4℃, to remove insoluble materials.

12. Collect supernatant carefully. Add 1mL 50% glutathione agarose beads suspension and mix gently over 2min at room temperature. Wash by adding 50mL ice-cold PBS and repeat two more times. Resuspend beads with ice-cold PBS in a small volume (1~2mL) and transfer to an Eppendorf tube.

13. Discard supernatant by short centrifuge, and collect the beads. Elute fusion protein by adding 1mL 50mmol/L Tris-Cl (pH8.0)/5mmol/L reduced glutathione. Mix gently for 2min. Centrifuge cells at 500g×10s and then collect supernatant. Repeat the wash 2~3 times and analyze each fraction by SDS-PAGE. Store eluted protein in aliquots containing 10% glycerol at -70℃. Determine the yield of fusion protein by measuring A_{280}. For GST protein, $A_{280} = 1$ corresponds to a protein concentration of 0.5mg/mL.

Part II Detection and Identification of GST Fusion Protein

Principle

Separation of protein by SDS-PAGE normally is one of simplest methods for detecting purity of protein product. In a denatured state, with strong reducing agents to remove secondary and tertiary structure, proteins are separated according to size. Hence, the purity of protein sample can be identified by the number of bands on the gel after staining.

Western blotting (alternately, immunoblotting) that combines electrophoresis and protein blotting, is a method of detecting specific protein in a crude extract or a more purified preparation. The first step uses gel electrophoresis (normally PAGE or SDS-PAGE) to separate native or denatured proteins. In order to make the proteins accessible to antibody detection, the proteins within the gel are then electrotransferred onto a membrane made of nitrocellulose (NC) or polyvinylidene fluoride (PVDF). Non-specific protein binding on the membrane is based upon hydrophobic interactions, as well as charged interactions. So after blocking non-specific binding on the membrane using bovine serum albumin (BSA) or non-fat milk, with a detergent such as Tween 20, antibodies (both monoclonal and polyclonal antibodies) only attach on the binding sites of the specific target protein. Last detection for the protein of interest is "probed" by a modified antibody which is linked to a reporter enzyme, such as alkaline phosphatase (AP) or horseradish peroxidase (HRP), which drives a colorimetric reaction or chemiluminescence reaction (Fig. 9-4).

Western blotting also allows investigators to measure relative amounts of the protein present in different samples. Normally the process is repeated for a structural protein, such as β-actin or tubulin, which should not change among samples, to control groups.

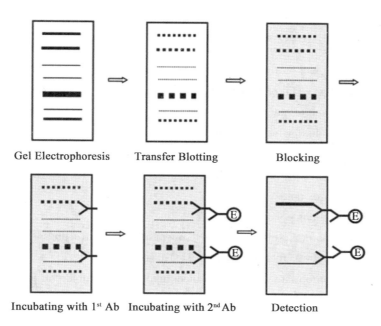

Fig. 9-4 The basic procedure of Western blotting

Reagents

1. Reagents for SDS-PAGE (see Experiment 13 Part II).
2. Prestaining protein standards.
3. 1×Transblotting buffer: dissolve 3.03g Tris base; 14.4g glycine into 200mL dH$_2$O, mix with 200mL methanol, add dH$_2$O up to 1 L. The pH should be 8.3 without adjustment.
4. TBS buffer: add 1.22g Tris (10mmol/L) and 8.78g NaCl (150mmol/L) to 1L dH$_2$O, and adjust pH to 7.5 with HCl.
5. TBS-T buffer: add 0.5mL Tween 20 (0.05%) to 1L TBS buffer.
6. Blocking buffer: dissolve 3.0g bovine serum albumin (BSA) or 5.0g non-fat milk into 100mL TBS-T buffer, keep at 4℃.
7. First (primary) antibody (1st Ab): rabbit anti-GST monoclonal Ab.
8. Second antibody (2nd Ab): goat anti-rabbit IgG (HRP labelled).
9. Developing reagent: dissolve DAB 6mg in 10mL TBS-T buffer, and mix with 0.05mL 30% H$_2$O$_2$. Enhanced chemiluminescent (ECL) kit can also be used.

Procedure

1. Seperation of protein by SDS-PAGE

(1) Preparation of samples:

① Cell lysates: collect cells and lyse the cells pellet with 100μL SDS-PAGE sample loading buffer. Boil for 5min, then cool at room temperature.

② Protein solution: after determine the protein concentration (Bradford assay, A_{280}, or BCA method), mix with equal volume 2× sample loading buffer, boil for 5min and cool.

(2) Loading and running the gel (see Experiment 5): after separation, take out the gel, cut a corner of gel for orientation and put gel in the transblotting buffer.

2. Electroblotting:

(1) Preparation of membrane: cut a piece of NC membrane according to the size of gel, and wet for about 10min in transblotting buffer.

(2) Assemble gel-membrane "sandwich": prewet the sponge pads, filter papers (slightly bigger than gel) in transblotting buffer. Put a sponge pad on the black piece of the transfer cassette. Arrange the sandwich according to following order: Black side-Sponge →Filter paper →Gel →NC membrane →Filter paper →Sponge-White side.

Close the sandwich with the white piece of the cassette and put in the transfer tank, put the black side near the black side of the device. Fill the buffer tank with the transfer buffer.

(3) Attach the electrodes. Turn on power supply to 100V (constant voltage) for 1h at 4℃. Bigger proteins might take longer to transfer. Using prestaining protein standards is easy for checking whether protein has been transferred from gel to membrane or not.

3. Blocking and Immunodetection:

(1) Disconnect transfer apparatus, take out the NC membrane, remove the membrane to a small container, immerse membrane in blocking buffer, rock gently for at least 1h.

(2) Add first antibody at appropriate dilution into blocking buffer, incubate for at least 1h at room temperature or overnight at 4℃.

(3) Pour off 1st antibody solution. Wash the membrane 3 × 10min with TBS-T buffer.

(4) Incubate with secondary antibody diluted in blocking buffer for at least 1h.

(5) Pour off 2nd antibody solution. Wash the membrane 3 × 10min with TBS-T buffer.

(6) Detected the specific protein band with enhanced chemiluminescent (ECL) kit or detected with colorimetric reaction: pour off TBS-T buffer, cover developing reagent to the membrane, monitoring development.

Notes

1. Acrylamide and bisacrylamide are neurotoxic and may be absorbed through the skin. Polyacrylamide is considered nontoxic but may contain unpolymerized material. When preparing and handling gels, you should wear gloves to avoid exposure to unpolymerized polyacrylamide.

2. One major difference between NC and PVDF membranes relates to the ability of each to support "stripping" off and reusing for subsequent antibody probes. The sturdier PVDF allows for easier stripping, and for more reuse before background noise limits experiments. Another difference is that PVDF must be soaked in 95% ethanol or methanol before use. PVDF membranes also tend to be thicker and more resistant to damage during use.

Experiment title

Date

Observations and results

Discussion

1. GST protein as a tag-protein is often used to expression and purification of foreign protein. With the use of _____ as a specific ligand for the GST protein, _____ is a simple and the most powerful technology for purification of GST fusion protein.

2. What is the principle and application of Western blotting? Why need "blocking" the NC membrane after electroblotting?

3. Describe the difference between NC and PVDF membrane.

Teacher's remarks

Signature

Date

Chapter 10 Isolation, Purification and Identification of Nucleic Acids

10.1 Preparation Principle and Procedure of Nucleic Acids

Isolation and purification of nucleic acids from various sources is the most important first step in molecular biology research, and the quality of nucleic acids product will be directly related to the success or failure of the further experiment.

Nucleic acids, including DNA and RNA, are mainly present as nucleoprotein complexes in cells. Eukaryotic chromosomal DNA exists as a double-stranded helical molecule. Chromosome, plasmid DNA in prokaryotes and mitochondria DNA in eukaryotic cells are double-stranded circular molecules. RNA exists in several different single-stranded structures with different characteristics, such as the poly A tail of eukaryotic mRNA. 95% of the eukaryotic DNA is in the nucleus, others are organelle DNA, such as mitochondria DNA, etc. 75% of RNA exists in the cytoplasm, and 10% in the nucleus. The others are in the organelles.

The criteria of isolation and purification of nucleic acids are the following: ①maintaining the integrity of primary structure, which is essential for further research of nucleic acids; ②removing contaminating materials such as proteins, lipids, polysaccharides and other high molecular weight compounds, including removing other types of nucleic acids molecules, for example, the contaminating DNA needs to be removed when isolating RNA.

To ensure the integrity and purity, the preferred method for isolation of nucleic acids from any source should be simple and rapid. Generally, the following issues should be considered: ①reduce chemical degradation of nucleic acids, avoiding the cleavage of phosphodiester bond by base and acid, solution of pH4~10 is common selected; ②reduce mechanical damage for high molecular weight nucleic acids, such as vortex, strong stirring, repeated freezing-thawing, etc. Small circular plasmid DNA and RNA are relatively resistant to mechanical damage; ③prevent hydrolysis of nucleases for nucleic acids. Adding EDTA and citrate to buffer can chelate Mg^{2+}, Ca^{2+} ions which are required for DNase activity. Especially in the process of RNA isolation, because of high activity of DNase, cell lysis buffer should contain RNase inhibitor such as bentonite, guanidine isothiocyanate. Fresh tissue or cell samples will be the best choice. Materials should be stored in liquid nitrogen or $-70°C$ if not used immediately.

The main steps involved in isolation of nucleic acids include cells disruption and removal of contamination of other DNA or RNA, protein, polysaccharides, lipids, other macromolecules. At last, salts and organic solvents should be discarded if nucleic acid molecules need to be purified.

10.1.1 Isolation of eukaryotic genomic DNA

DNA is very inert molecules with double strands closely linked by hydrogen bonds. In addition, bases are protected by lateral phosphate and ribose, which strengthened the internal stacking force. However, high molecular weight DNA is still fragile. Long and curved DNA molecules in the solution are easy to be broken by mechanical damage such as fluid suction, vortex, and strong stirring. Therefore, it is difficult to obtain larger molecular weight

DNA, especially for more than 150kb.

All nucleated cells in eukaryotes (including culture cells) can be used for genomic DNA preparation. The isolation protocols consist of two parts—gently lyse the cells and solubilize the DNA, followed by several enzymatic or chemical methods to remove contaminating RNA, proteins, polysaccharides, and other macromolecules.

Various methods used to break eukaryotic cells include ultrasonication, homogenation, and low osmotic pressure, etc. Gentle detergent or proteinase K treatment is usually used to obtain high molecular weight DNA. Proteins are removed by treatment with phenol or mixture of chloroform-isoamyl alcohol or phenol/chloroform. Repeated extraction operation may damage the DNA molecules by shearing force. So using 80% formamide plus dialysis method can obtain DNA fragments up to 200kb. It depends on different requirements to select different isolation methods.

10.1.2 Isolation of plasmid DNA

A plasmid is an extra-chromosomal, circular DNA molecule that replicates independently of the chromosomal DNA of bacteria. It is much smaller than chromosomal DNA (MW range from 1kb to 200kb). The replication of plasmid includes two types: stringent control and relaxed control. The former is low copy number plasmids. The latter is high copy number plasmids (usually contains more than >100 copies).

Plasmids can carry foreign DNA into bacteria for amplification or expression, so it is called vectors in genetic engineering. Many plasmids (pUC18, pBR322, etc.) are commercially available. A typical plasmid (vectors) used must contain an origin (*Ori.*) of replication, a region called multiple clone site (MCS) for the insertion of the target DNA, and contain a selectable marker such as a gene for antibiotic resistance (e.g., ampicillin).

The isolation and extraction of plasmid DNA are the most popular techniques in molecular cloning. The isolation of plasmid normally contains three steps: growth of bacteria containing the plasmid, separation and lysis of the bacteria from the culture medium, and purification of the plasmid DNA.

Methods for preparation of plasmid DNA include SDS-alkaline denaturation, salt-SDS precipitation and rapid boiling, etc. The SDS-alkaline denaturation method is a popular procedure for plasmid preparation because of its overall versatility and consistency. The gradients centrifugation in CsCl-ethidium bromide is useful for large-scale preparation of macro, closed plasmid DNA. But it is expensive and time-consuming, and has been substituted by ion exchange chromatography, gel filtration now. Plasmid purification kits including spin chromatographic column are also commonly used. Plasmid DNA purified by above-mentioned methods can be used for PCR, transformation, restriction enzyme analysis, DNA cloning, and probe labelling.

10.1.3 Isolation of tissue/cell RNA

Isolation of pure and intact RNA from cells is crucial for many molecular biological experiments, and is the basis of gene expression analysis. For example, the successful performing of cDNA synthesis, Northern blotting and in vitro translation experiments are extremely determined by the quality of RNA.

A typical mammalian cell contains about 10^{-5} μg RNA, of which 80%~85% is rRNA (mainly including 28S, 18S, 5.8S, and 5S), and only 1%~5% is mRNA. According to their density and molecular size, separation methods involve density gradient centrifugation, gel electrophoresis, anion-exchange chromatography and HPLC. Various mRNA which function as genetic messengers, is one of the main research subjects in molecular biology. Eukaryotic mRNA molecules have a wide range of molecular sizes and different nucleotide sequences, but with 3'-polyA tail (20~250 nucleotides in length), therefore, mRNA can be separated using oligo (dT)-cellulose chromatography.

For a wide range of needs and applications in molecular biology, many simple and high-yielding methods have been developed recently for isolation and purification of total RNA, mRNA and specific RNA.

RNA is more active than DNA because of the 2′, 3′ hydroxyl residues of ribose. RNA is high sensitive to RNA enzyme (RNase), which has very stable activity, particularly pancreatic RNase. RNase exists not only in cells, but also in dust, various containers and test reagents, human sweat and saliva. It is heat, acid and alkali resistant, and its activity is not influenced by divalent metal ion chelators. Therefore, the key for RNA isolation is controlling the contaminating RNase, including exogenous RNase and endogenous RNase.

10.1.3.1 Prevention of contamination with exogenous RNase

Exogenous RNase mainly comes from operators' hands, saliva, and cells, mycetes in the dust. Therefore, disposable masks and gloves should be worn, and clean environment is also necessary. All operations should be conducted on ice, as low temperature can reduce the activity of RNase.

All glasswares should be washed and roasted at 180℃ for 3h before use, or rinsed with 0.1% diethylpyrocarbonate (DEPC) to inactivate RNase and then autoclaved to evaporate DEPC. It is the best to choose the disposable RNase-free plastics, such as the Eppendorf tubes, tips. Electrophoretic tanks should be soaked in 3% H_2O_2 for 10min, then rinse with DEPC treated water thoroughly. It is better to prepare special tank for RNA electrophoresis.

All solutions should be added DEPC to 0.05% ~ 0.1%, left overnight at room temperature, or stirred by magnetic force for 20min before autoclave (15Ibf/in^2, 15 ~ 30min). Some solutions which cannot be autoclaved, should be treated with DEPC-treated ddH_2O, and sterilized through 0.22μm membrane filter. DEPC will react with Tris base rapidly (only 1.25min of half-life), and Tris buffers should be made up of sterilized DEPC-treated ddH_2O. Phenol for RNA extracting protein during RNA isolation should be prepared separately, and added 8-hydroxyquinoline to 0.1%, which can inhibit RNase and oxidant.

10.1.3.2 Inhibiting the activity of endogenous RNase

Release of endogenous RNase from tissues or cells is the major cause of RNA degradation during cell lysis. It is very important to inhibit the RNase activity as early as possible by removing cell protein and adding RNase inhibitor such as guanidine isothiocyanate.

The amount of endogenous RNase differs in tissues. Pancreas and spleen cells are rich in RNase, and it is essential to use several strong RNase inhibitors.

10.2 Identification and Analysis of Nucleic Acids

Preparation of high-MW DNA/RNA with high purity is the basic requirement for further research of gene structure and function. Commonly used methods for identification and analysis of nucleic acids include UV spectrophotometry, gel electrophoresis, hybridization, PCR and DNA sequencing, etc.

10.2.1 Spectrophotometry

Both purine and pyrimidine bases of nucleic acids have the maximum absorption peak at 260nm wavelength, which can be used for quantitative analysis of nucleic acids. UV spectrophotometry is only used to determine the concentration of nucleic acid solutions more than 0.25μg/mL. For a quite diluted solution, fluorescence spectrometry can be used to estimate the concentration of nucleic acids. The absorption of 1OD (A) is equivalent to approximately 50μg/mL dsDNA, 33μg/mL single strand DNA (ssDNA), or approximately 40μg/mL RNA. The ratio A_{260}/A_{280} is used to estimate the purity of nucleic acid, since proteins absorb ultraviolet(UV) light at 280nm. Pure DNA should have a ratio of approximately 1.8, whereas pure RNA should give a value of approximately 2.0. Absorption at 230nm reflects contamination of substances such as carbohydrates, peptides, phenols or aromatic compounds. In the case of pure samples, the ratio A_{260}/A_{230} should be approximately 2.2.

10.2.2 Gel electrophoresis of nucleic acids

Nucleic acids are uniformly negatively charged, and will migrate to the anode in electric field. Gel electrophoresis separations can be used either for preparation or for analysis of nucleic acids. DNA fragments with specific molecular weight (MW) can be recovered and purified after separation by gel electrophoresis. Nucleic acid molecules have similar charge density. Gel with appropriate concentration is selected as the electrophoretic support medium, which has molecular sieve effect, can separate nucleic acid molecules based on different molecular size and conformation.

10.2.2.1 Factors affecting DNA electrophoresis

1. The nature of nucleic acids

Charge, molecular size and conformation of nucleic acids determine the electrophoretic mobility. Generally, linear dsDNA don't have complex conformation affecting the mobility, and the relationship between relative mobility and logarithm of MW is inversely proportional in the gel electrophoresis. Molecular conformation can also affect the mobility. For example, the mobility of plasmid DNA with the same MW is: closed-loop > linear > open-loop. The mobility of ssDNA or RNA is affected by base composition and molecular size in gel electrophoresis. Therefore, single-stranded nucleic acids can be separated using denaturing gel electrophoresis.

2. Pore size of gel

Agarose and polyacrylamide gel (PAG) are supporting media of gel electrophoresis, often used for nucleic acid separation. Changes of gel concentration can adjust the pore size of the gel. Agarose gels have larger pore size and lower resolution, but have a wide range to resolve DNA fragments from 100bp to 50kb. DNA fragments less than 20kb is mostly suitable to horizontal gel electrophoresis separation in the electric field with constant intensity and direction. Larger DNA molecules require some special pulsed-field gel electrophoresis (PFGE) to be separated. PAG can be used to separate DNA fragments from 5bp to 500bp, and can identify the difference of 1bp between DNA fragments. Table 10.1 lists different gel concentration with their corresponding separation range of linear DNA.

Table 10.1　**Gel concentration and effective separation range of linear DNA**

Gel	Gel concentration(%)	Separation range of linear DNA (bp)	Bromphenol blue (equivalent to DNA) (bp)
Agarose	0.3	5,000~60,000	
	0.7	800~10,000	1,000
	0.9	500~7,000	600
	1.2	400~6,000	
	1.5	200~4,000	
	2.0	100~3,000	150
PAG	3.5	100~2,000	100
	5.0	80~500	65
	8.0	60~400	45
	12.0	40~200	20
	15.0	25~150	15
	20.0	6~100	12

3. Electric field strength and direction of the electric field

The mobility of linear DNA fragments is directly proportional to the voltage when voltage is low. Usually the electric strength is not more than 5V/cm gel. If electric field intensity increases, migration of DNA fragments with high MW will be increased in different range, and the effective separation range of the gel will reduce, 0.5~1.0V/cm overnight electrophoresis is often used to separate DNA fragments with high MW in order to get higher resolution and neat bands.

4. Electrophoretic buffer

Buffer composition and ionic strength directly affect electrophoretic mobility. Gels are usually run in Tris-acetate-EDTA (TAE), Tris-borate-EDTA (TBE) or Tris-phosphate-EDTA (TPE) buffer. TAE is used traditionally, but its buffer capacity is very low. TBE and TPE have higher buffer capacity, but borate can form complexes with agarose in TBE buffer. So in a short period of agarose gel electrophoresis ($\leqslant$5h), TAE is still used, and TBE is often used in PAGE.

10.2.2.2 Tracker and dye in gel electrophoresis of nucleic acids

1. Tracker

Colored markers are usually used in the sample buffer during electrophoresis to indicate the migration of samples, which include bromphenol blue (BB) and green xylene (blue). Molecular weight of BB is 670U. The mobility of BB is almost equivalent to 1kb and 0.6kb, 0.15kb of linear dsDNA in 0.6%, 1%, 2% agarose gel electrophoresis, respectively. The MW of green xylene is 554.6U, and the quantity of electricity is less than BB, which results in slower mobility. The mobility of green xylene is equivalent to 4kb linear dsDNA fragment in 1.0% agarose gel electrophoresis.

2. Dye

After electrophoresis, nucleic acids need to be stained to show the bands. The following are dyes for nucleic acids usually used:

(1) Ethidium bromide (EB): it is the most common fluorescence dye for nucleic acids, which can embed into dsDNA base pairs, and issue orange fluorescence in the ultraviolet excitation. The fluorescence intensity of EB-DNA complexes is 10 times stronger than free EB in gels. Even 10ng DNA samples can be detected. EB can also be used to detect ssDNA/RNA, but the fluorescence intensity is relatively low because of lower affinity with ssDNA/RNA.

The final concentration of 0.5μg/mL EB in the gel can be observed the migration of nucleic acids at any time. However, EB with positive charge will increase the length of open-loop and linear DNA molecules, and make them more rigid after inserting into bases, and the mobility of linear DNA decreases by 15 percent rate. In addition, the gel can be immersed in 0.5μg/mL of EB solution for 10min after electrophoresis if it is used for measure the MW of nucleic acid. EB is easily decomposed in the light, and should be kept dark at 4℃.

(2) SYBR Green: it is an asymmetric cyanine compound with high affinity with nucleic acids, and it is a kind of fluorescent dyes with highly sensitive introduced by Molecular Probes Incorporated in 1995. The resulting nucleic acids-dye-complex absorbs blue light (λ_{max} = 497nm) and emits green light (λ_{max} = 520nm). SYBR series fluorescent dyes have been highly praised and widely used in molecular biology laboratories in recent years, including SYBR Green I, SYBR Green II and SYBR Gold.

SYBR Green I preferentially binds to dsDNA, but stains single-stranded nucleic acids with lower performance. It emits bright green under the excitation of ultraviolet light, and its fluorescence intensity is greatly enhanced when it binds to DNA (the fluorescence intensity is at least 10 times stronger than that of EB). SYBR Green II is widely used in the detection of RNA or ssDNA staining in gel electrophoresis. It emits bright yellow-green under UV irradiation (slightly yellower than SYBR Green I). SYBR Gold is later developed which can be used for the detection of dsDNA, ssDNA and RNA, and it has higher sensitivity. SYBR Gold emits bright golden yellow under ultraviolet light.

It can be used in quantitative PCR system, various electrophoresis (such as PFGE, capillary electrophoresis, PAGE, agarose gel electrophoresis) and in situ staining. SYBR fluorescent dye does not affect the activity of most enzymes in the experiments of molecular biology, such as *Taq* DNA polymerase, reverse transcriptase, restriction endonuclease, T4 DNA ligase. It is simple and convenient to use, does not require dyeing and cleaning process.

Any small molecule capable of binding DNA with high affinity is a possible carcinogen, including SYBR Green. According to the ability of chemicals to cause mutations, the mutagenicity of SYBR dyes is much smaller than that of EB.

(3) Ag^+: it can form stable complexes with nucleic acid, and then be reduced to silver particles by formaldehyde. $AgNO_3$ reagents can make DNA/RNA dark brown in PAG. The sensitivity of silver staining is about 200-fold higher than EB staining. It can detect RNA of 0.5ng in gel. But Ag^+ can also bind with protein, detergent and make them brown. Ag^+ binds with DNA stably and damages DNA, so silver staining is not suitable for the recovery of DNA fragments.

10.2.3 Nucleic acids hybridization

Identification of some specified fragments from complex mixtures separated by gel electrophoresis usually uses nucleic acids hybridization (such as Southern blotting and Northern blotting). The common procedure of nucleic acids hybridization includes: separating nucleic acids by gel electrophoresis, transferring all fragments to nylon membrane or nitrate cellulose (NC) membrane, and then using labelled probe to hybridize with target fragments for identification.

10.2.4 Polymerase chain reaction (PCR)

PCR is a rapid procedure for in vitro enzymatic amplification of specific DNA fragments, which for gene amplification, separation, screening, sequence analysis or identification. PCR involves sequential cycles composed of denaturation, annealing, extension three steps. There have been numerous applications in molecular cloning, diagnosis of genetic diseases, and forensic science, etc.

10.2.5 DNA sequencing

DNA sequencing is mainly used for identification of new cDNA cloning, DNA mutation, linkage, accuracy of PCR products, and analysis of non-coding gene sequences. At present, automated DNA sequencing has been widely used in molecular biology research.

10.2.6 RNA analysis

One of the main uses of RNA isolation is to analyze gene expression. Commonly used methods for analysis of RNA structure and quantity involve SI endonuclease analysis, ribonuclease protection experiments, primer extension, and Northern blotting. Nuclear run-off transcription analysis technique can be used to determine the number of activated RNA polymerase in a known gene in eukaryotic cells.

Experiment 13 Isolation and Identification of Eukaryotic Genomic DNA

Part I Isolation of Eukaryotic Genomic DNA

Protocol I Isolation of Genomic DNA by Proteinase K-Phenol Method

Principle

Genomic DNA molecules exist as nucleoprotein in cells. The isolation principle of DNA is not only to separate DNA from complex with proteins, lipids and sugars, but also to maintain its integrity. This is the method of choice when high-molecular-weight DNA is required. Cells or tissues are lysed with SDS and proteinase K. EDTA chelates Ca^{2+} and Mg^{2+} to inhibit degradation of DNA by DNase. Phenol and chloroform/isoamyl alcohol extractions follow to remove most of the non-nucleic acid organic molecules (isoamyl alcohol gets rid of the foam). The DNA is precipitated out of solution with 70% ethanol, and redissolved in TE buffer (pH8.0). Approximately 200μg of eukaryotic DNA, 100~150kb in length, is obtained from 5×10^7 cultured mammalian cells. DNA obtained can be used for Southern blotting, PCR or for construction of genomic libraries.

Reagents

1. Mammalian cell lysis buffer: 10mmol/L Tris-Cl; 0.1mol/L EDTA, pH8.0; 0.5% SDS; 20μg/mL RNase A (DNase free).
2. Proteinase K (20mg/mL).
3. Phenol, equilibrated with 0.5mol/L Tris-Cl (pH8.0).
4. chloroform/isoamyl alcohol (24 : 1, $V : V$).
5. Sodium acetate (3mol/L, pH5.2): dissolve 40.82g NaAc ($3H_2O$) in 60mL ddH_2O. Add enough glacial acetic acid to adjust pH to 5.2, and add water to 100mL. Store at 4℃ after autoclave.
6. 100% ethanol.
7. 70% ethanol.
8. 0.1mol/L Tris-Cl (pH8.0).
9. TE buffer (10mmol/L Tris-Cl, 1mmol/L EDTA, pH8.0).
10. TBS (Tris-buffered saline, pH7.4): dissolve 8g NaCl, 0.2g KCl and 3g Tris in 800mL ddH_2O, add HCl to pH7.4, then add ddH_2O to 1L.

Procedure

1. According to the different samples, choose one of methods as step 1.
(1) Cell samples.
① Lyse Cells growing in monolayer cultures.

Take one batch of culture dishes, immediately remove the medium by aspiration. Wash the monolayers of cells twice with ice-cold TBS. Scrape the cells into 1mL TBS, and then transfer the cell suspension to a centrifuge tube on ice. Recover the cells by centrifugation at 1500g for 10min at 4℃. Resuspend the cells in 5~10 volumes of ice-cold TBS and repeat the centrifugation. Resuspend the cells in TE buffer at a concentration of 5×10^7cells/mL. Add 1mL lysis buffer per 0.1mL cell suspension. Incubate the suspension for 1h at 37℃, and then proceed immediately to step 2.

② Lyse cells growing in suspension cultures.

Transfer the cells to a centifuge tube and recover them by centrifugation at 1500g for 10min at 4℃. Remove the supernatant. Wash the cells with ice-cold TBS and centrifuge repeat twice. Remove the supernatant and gently suspend the cells in TE at a concentration of $5×10^7$ cells/mL. Add 1mL lysis buffer per 0.1mL cell suspension. Incubate the solution for 1h at 37℃ and then proceed immediately to step 2.

(2) Tissue samples.

Drop approx. 1g of freshly excised tissue into liquid nitrogen in the stainless-steel container. Blend at top speed until the tissue is ground to a powder. Allow the liquid nitrogen to evaporate, and add the powdered tissue little by little to approx. 10 volumes (W/V) of lysis buffer in a beaker. Allow the powder to spread over the surface of the lysis buffer, and then shake the beaker to submerge the material. Transfer the suspension to a 50mL centrifuge tube. Incubate the tube for 1h at 37°C, and then proceed to step 2.

(3) Blood samples.

① To collect cells from freshly drawn blood.

Collect approx. 2mL of fresh blood in tubes containing either ACD (citric acid 0.48%, sodium citrate 1.32%, dextrose 1.47%) or EDTA. Transfer the blood to a centrifuge tube and centrifuge at 1300g for 15min at 4℃. Remove the supernatant. Use a Pasteur pipette to transfer the buffy coat carefully to a fresh tube and repeat the centrifugation. Discard the pellet of red cells. Remove residual supernatant and resuspend the buffy coat in 1.5mL lysis buffer. Incubate the solution for 1h at 37℃, and proceed to step 2.

② To collect cells from frozen blood samples.

Thaw the blood in a water bath at room temperature. Add an equal volume of phosphate-buffered saline and centrifuge the blood at 3500g for 15min at room temperature. Remove the supernatant, which contains lysed red cells. Resuspend the pellet in 1.5mL of lysis buffer. Incubate the solution for 1h at 37℃, and then proceed to step 2.

2. Add proteinase K (20mg/mL) to a final concentration of 100μg/mL. Use a glass rod to mix the solution gently into the viscous lysate of cells. Incubate the lysate in a water bath for 1h at 37℃ followed by 3h at 50℃. Swirl the viscous solution from time to time.

3. Cool the solution to room temperature and add an equal volume of phenol equilibrated with 0.1mol/L TrisCl (pH8.0). Gently mix the two phases by slowly turning the tube end-over-end for 10min on a tube mixer or roller apparatus.

4. Separate the two phases by centrifugation at 5,000g for 15min at room temperature. Transfer the viscous aqueous phase to a fresh centrifuge tube. Repeat the extraction with phenol twice more if it is necessary.

5. Add an equal volume of chloroform/isoamyl alcohol (24:1), place on slowly rotating wheel for 15min. Centrifuge 5,000g for 10min. Transfer the pooled aqueous phases to a fresh centrifuge tube and add 0.1 volume of 3mol/L sodium acetate. Add 2.5 volumes of ice-cold anhydrous ethanol and swirl the tube until the solution is thoroughly mixed. The DNA immediately forms a precipitate.

6. Centrifuge 10,000g for 15min. Discard the supernatant. Wash the DNA precipitate twice with ice-cold 70% ethanol, and collect the DNA by centrifugation at 5,000g for 5min.

7. Remove as much of the 70% ethanol as possible. Store the pellet of DNA in an open tube at room temperature until the last visible traces of ethanol have evaporated. Add 0.1mL of TE (pH8.0) and place the tube on a rocking platform and gently rock the solution for 12~24h at 4℃ until the DNA has completely dissolved. Store the DNA solution at 4℃.

8. Analyze the quality of the preparation of high-MW DNA by PFGE or by electrophoresis through 0.8% agarose gel. Purity of DNA product and the concentration of DNA could be detected by UV spectrophotometry.

Notes

1. All glasswares and solutions (except organic solvents) should be sterilized. Gloves should be worn to avoid contamination of the experimental material and apparatus with nucleases which occur in fair abundance in skin exudates.

2. The pH of the phenol must be approx. 8.0 to prevent DNA from becoming trapped at the interface between the organic and aqueous phases.

3. When transferring the aqueous (upper) phase, it is essential to draw the DNA into the pipette very slowly to avoid disturbing the material at the interface and to minimize hydrodynamic shearing forces.

4. Do not allow the pellet of DNA to dry completely; desiccated DNA is very difficult to dissolve.

Protocol II Preparation of Genomic DNA by Adsorption Column

Principle

After genomic DNA is extracted from cells of tissue, it also can be prepared by adsorption Column, which is supported by Genomic DNA Preparation Kit. The kit employs a special lysis buffer and proteinase K to efficiently release genomic DNA from materials, then precipitate the contaminating proteins, carbohydrates and lipids to efficiently segregated genomic DNA. The genomic DNA can be selectively adsorbed by special adsorption column. After washing and desalting, the purified DNA is eluted in either buffer or ddH_2O. The adsorption column can bind and purify up to 20μg of genomic DNA, and the DNA is suitable for a variety of applications in molecular biology, such as PCR, Southern blotting, etc.

Reagents

1. Mammalian cell lysis buffer: 10mmol/L Tris-Cl; 0.1mol/L EDTA, pH8.0; 0.5% SDS; 20μg/mL RNase A (DNase free).

2. Proteinase K (20mg/mL).

3. Protein precipitation buffer.

4. Buffer W1: wash buffer.

5. Buffer W2 concentrate desalting buffer: add marked amount of ethanol to the buffer W2 concentrate before using.

6. Eluent buffer: 2.5 mM Tris-Cl, pH8.5.

7. Adsorption column set including adsorption column, 2mL microfuge tube, 1.5mL Ep tube.

Procedure

1. Lysis and homogenization (refer to Protocol I)

2. Collect 350μL of the homogenate and transfer to a new Ep tube.

3. Add 20μL of Proteinase K and 150μL of Mammalian cell lysis buffer, mix immediately, lysis the cells sufficiently, vortex if necessary. Incubate at 56℃ for 15min.

4. Add 350μL of Protein precipitation buffer and mix well immediately.

5. Centrifuge at 10,000g for 10min at room temperature to pellet cellular debirs.

6. Place adsorption column into a 2mL microfuge tube (provided by kit). Pipette the supernatant into the column. Centrifuge for 1min at 12,000g.

7. Discard the filtrate from the 2mL microfuge tube. Place the column back into the tube. Pipette 500μL of buffer W1 to the column and centrifuge at 12,000g for 1min.

8. Discard the filtrate and place the column back into the 2mL tube, add 700μL of buffer W2 and centrifuge at 12,000g for 1min.

9. Discard the filtrate and repeat this wash step with 200μL of buffer W2.

10. Discard the filtrate, place the column back into the 2mL tube and centrifuge at 12,000g for 1min.

11. Transfer the adsorption column into a new 1.5mL Ep tube (provided by kit). Elute the genomic DNA by adding 100~200μL Eluent buffer (or deionized water) to the center of the adsorption membrane. Let it stand for 1min at room temperature, centrifuge at 12,000g for 1min. The genomic DNA was collected in the Ep tube.

Notes

1. Check cell lysis buffer and protein precipitation buffer before each use. If precipitation occurs in the buffer, incubate at 37℃ to dissolve the precipitate.

2. Pre-warming the eluent to 65℃ to improve elution efficiency.

3. Do not add proteinase K directly to cell lysis buffer.

Protocol III Preparation of White Blood Cell DNA by NaI Method

Principle

Nucleated blood cells are used to prepare genomic DNA. In the extraction solution, high concentration of NaI serves as detergent to dissolve nuclear membrane and separate DNA from histone. EDTA chelates Ca^{2+} and Mg^{2+} to inhibit degradation of DNA by DNase. Then chloroform/isoamyl alcohol precipitate proteins and dissolve lipids (isoamyl alcohol get rid of the foam). Isopropanol may be used to precipitate DNA in the water phase. Finally, DNA pellet is washed by 37% isopropanol or 70% ethanol, then dissolve in water or TE buffer (pH8.0).

Reagents

1. 1×TE buffer (pH8.0): 10mmol/L Tris-Cl, 1mmol/L EDTA-Na_2, sterilized.

2. 10% SDS.

3. 6mol/L NaI (sodium iodide).

4. Chloroform/isoamyl alcohol (24 : 1, V/V).

5. Isopropanol.

6. 37% isopropanol.

7. 70% ethanol.

8. double distilled water (sterilized) (ddH_2O).

Procedure

1. Pipette 0.1mL peripheral blood to a 1.5mL Eppendorf (Ep) tube, then centrifuge 10,000r/min for 1min. Discard the upper aqueous layer and add 200μL ddH_2O. Mix gently for 20s.

2. Add 200μL NaI, mix it by hand with up-down motion for 20s.

3. Add 400μL Chloroform/isoamyl alcohol. Mix it by hand thoroughly for 20s, and then centrifuge at 12,000 r/min for 12min.

4. Remove 360μL of upper aqueous layer carefully with an adjustable pipette and transfer it to a new 1.5mL Ep tube. Add 200μL isopropanol. Mix up by hand for 20s. After 15min at room temperature, centrifuge at 14,000r/min at 4℃ for 12min.

5. Discard the supernatant carefully and wash the pellet by adding 1mL of 37% isopropanol or 70% ethanol (don't vortex). Centrifuge 12min at 14,000r/min, discard the supernatant carefully.

6. Remove the last trace liquid on the inner wall of tube by pipette or filter paper. Evaporate the isopropanol at room temperature for 15min.

7. Add 50μL TE buffer to dissolve the pellet gently. Employ agarose gel electrophoresis to determine the yield and purity of DNA isolation or store the samples at −20℃.

Notes

1. Blood samples must be fresh. All performance should be done at 4℃.
2. The Ep tubes and tips must be sterilized.
3. Severely shaking will break DNA chain, so shaking should be gently either by hand or by vortexer.

Part II Identification of Genomic DNA by UV Spectrophotometry and Gel Electrophoresis

Protocol I Identification of DNA by UV Spectrophotometry

With the use of UV spectrophotometry, the quantitative analysis of nucleic acids has established as a routine method. Aqueous buffers with low ion concentrations (e.g., TE buffer) are ideal for this method. The DNA concentration is determined by measuring A_{260} and then calculating the concentration of the measured sample using according factors. The absorption of 1 A is equivalent to approximately 50μg/mL dsDNA or approximately 30μg/mL for oligonucleotides. Purity determination of DNA can be recognized by the calculation of "ratio". In the case of pure samples, the ratio A_{260}/A_{280} should be approximately 1.8, and the ratio A_{260}/A_{230} should be approximately 2.2. Phenol has an absorbance maximum of 270nm but the absorbance spectrum overlaps considerably with that of nucleic acids. If there is phenol contamination in your DNA sample, the absorbance at 260nm will be high, giving a false measure of DNA concentration.

You can calculate the concentration of the DNA in your sample as the following:

$$\text{DNA concentration (μg/mL)} = A_{260} \cdot (\text{dilution factor}) \cdot 50$$

Reagents

1. Distilled H_2O (dH_2O).
2. DNA solution (unknown concentration).

Procedure

1. Turn on machine using the switch in the back of the machine and wait for it to warm up. Open the lid on the top of the machine. Place the blank cuvette in the cuvette chamber in front of the light source, then set wavelength to 260nm. Wash quartz cuvettes thoroughly before use and if two cuvettes are used then make sure they are a matching pair.

2. Insert a cuvette within 1mL of water into machine and press "autozero".

3. Put 5μL of DNA with unknown concentration into a clean matching cuvette, add 1mL of water and mix by covering with parafilm.

4. Insert cuvette into machine and note reading, i.e., 0.30.

5. Follow step 3 again, but this time set wavelength to 280nm, note reading.

Take the A_{260} readings and multiply the figure by 10 to give the concentration of DNA per mL, i.e., 0.30× 10 = 3.0μg/μL.

Take A_{280} readings, divide A_{260} by the A_{280}, and this should give a figure around 1.8. If it is less than 1.7, the DNA is not pure and sample should be purified using phenol/chloroform.

Protocol II Identification of DNA by Agarose Electrophoresis

Agarose gel electrophoresis is employed to check the progression of a restriction enzyme digestion, to quickly determine the yield and purity of a DNA isolation or PCR product, and to size fractionate DNA fragments, which then could be eluted from the gel. Agarose powder can be heated and dissolved in a buffer, and the warm gel solution is then poured into a mold, which is fitted with a comb. The percentage of agarose in the gel varied and gels should be prepared depending on the size(s) of DNA fragment(s).

Agarose gels usually run in a horizontal configuration. Electrophoresis is usually performed at 150~200mA for 0.5~1h at room temperature. DNA markers are co-electrophoresed with DNA samples, when appropriate for estimating fragment size.

Nucleic acids are typically visualized by staining with EB or SYBR dyes. These dyes are included in the gel matrix to enable fluorescent visualization of the DNA fragments under UV light.

Reagents

1. 5×TBE buffer (89mmol/L Tris, 89mmol/L boric acid, 2mmol/L EDTA, pH8.3).
2. 6×DNA loading buffer (30% glycerol, 0.25% bromophenol blue, 0.25% green xylene).
3. 5mg/mL EB or SYBR Green I.
4. 1.2% agarose gel: add agarose 1.2g and 0.5×TBE 100mL into 0.5 L flask, and heat them in a microwave for 2~4min until the agarose is dissolved.
5. DNA markers.

Procedure

1. Assemble the gel casting tray and comb. The comb should not touch the bottom of the tray.
2. Melt the agarose gels repeatedly using a microwave. When gel solution is cooled to about 50~60℃, pour solution into the casting tray to ensure that there are no bubbles in the gel. Then, allow the gel to become solidify and slightly opaque within 20~30min.
3. Carefully remove the comb by lifting gently, and put the gel tray into the electrophoretic tank. Adding 0.5 ×TBE buffer to cover the gel by about 3~5mm. Ensure that the wells are submerged and filled with buffer.
4. Prepare the DNA samples for loading by mixing with DNA loading buffer. Pipette 10μL of each sample into individual wells with a gel loading tip.
5. Once all the samples are loaded, cap the lid of tank and connect the electrodes (the black wire represents negative electrode). Turn on the power supply to run at a constant voltage of 100~120V. Caution: Do not remove the lid from the apparatus without disconnecting the power wire.
6. Turn off the power supply and disconnect the power leads when the blue tracking dye has migrated about 75% of the distance to the end of the gel (usually within 60~90min).
7. Carefully transfer the gel into a plastic dish containing EB or SYBR Green I staining solution. Set in a dark drawer for 30min. Then DNA fragments are observed in a gel by UV detector.

Notes

1. When preparing agarose for electrophoresis, it is best to sprinkle the agarose into room-temperature buffer, swirl, and let sit at least 1min before microwaving. This allows the agarose to hydrate first, which minimizes foaming during heating.
2. The minimum amount of DNA detectable by EB on a 3mm thick gel and a 5mm wide lane is 1ng. Loading DNA in the smallest volume possible will result in sharper bands.

3. Migration of DNA is retarded and band distortion can occur when too much buffer covers the gel. The slower migration results from a reduced voltage gradient across the gel.

4. You can preserve DNA in agarose gels for long-term storage using 70% ethanol.

Experiment title

Date
Observations and results

Discussion

    1. Genomic DNA molecules exist as _____ in cells.

    2. On the process of isolating genomic DNA, the role of chloroform/isoamyl alcohol is _____ . The role of 37% isopropanol or 70% ethanol is _____ .

    3. EDTA can inhibit degradation of DNA by DNase through _____ .

    4. In Protocol II, high concentration of NaI serves as _____ and separate DNA from histone.

    5. The absorption of 1 OD (A) is equivalent to approximately _____ dsDNA. The ratio A_{260}/A_{280} is used to estimate the purity of nucleic acid. Pure DNA should have a ratio of approximately _____ .

    6. What is the basic principle of DNA isolation?

Teacher's remarks
Signature
Date

Experiment 14　SDS-Alkaline Lysis Preparation of Plasmid DNA

Principle

This method exploits the difference in denaturation and renaturation characteristics of covalently closed circular plasmid DNA and chromosomal DNA. Under alkaline conditions (at pH11~12), both plasmid and chromosomal DNA are efficiently denatured. Rapid neutralization with a high-salt buffer such as potassium acetate in the presence of SDS has two effects. (1) Rapid neutralization causes the chromosomal DNA to base-pair in an intrastrand manner, forming an insoluble aggregate that precipitates out of solution. The covalently closed nature of the circular plasmid DNA promotes interstrand base-pairing, allowing the plasmid to remain in solution. (2) The potassium salt of SDS is insoluble, so the protein precipitates and aggregates, which assists in the entrapment of the high-molecular-weight chromosomal DNA. Separation of soluble and insoluble material is accomplished by a clearing method (e. g., filtration, magnetic clearing or centrifugation). The soluble plasmid DNA is ready to be further purified.

Reagents

1. Solution Ⅰ: 25mmol/L Tris-Cl, 10mmol/L EDTA, 50mmol/L glucose (pH8.0).

Add 1.0mol/L Tris-Cl (pH8.0) 1250μL, 0.5mol/L EDTA (pH8.0) 1mL, and glucose 450mg into 30mL ddH$_2$O. Bring final volume to 50mL. Store at 4℃ after autoclave.

2. Solution Ⅱ: 0.2mol/L NaOH, 1.0% SDS (freshly prepared).

Add 1.0mol/L NaOH 20mL, 20% SDS 5mL into 50mL ddH$_2$O. Bring final volume to 100mL. Store at room temperature.

3. Solution Ⅲ: 3mol/L potassium acetate, pH4.8.

Add 5mol/L potassium acetate 60mL into 10mL ddH$_2$O. Add enough glacial acetic acid to bring pH to 4.8 (approx. 11mL). Bring final volume to 100mL.

4. 3mol/L Sodium acetate, pH5.2 adjusted by acetic acid.

5. Phenol, equilibrated with 0.5mol/L Tris-Cl (pH8.0).

6. Chloroform/ Isoamyl alcohol (24 : 1, $V:V$).

7. 70% Ethanol.

8. TE buffer (10mmol/L Tris-Cl, 1mmol/L EDTA, pH8.0).

Procedure

1. Inoculate 5mL of LB medium supplemented with the appropriate antibiotic with a single colony from a plate culture. Grow overnight at 37℃ with shaking.

2. Centrifuge 1.0mL of the culture at full speed in an Eppendorf (Ep) tube to pellet the bacteria, discard the supernatant. (Select to repeat the procedure by adding an additional 1.0mL from the same culture, and centrifuge to get more bacterial precipitation.)

3. Resuspend the bacterial precipitation with 100μL Solution Ⅰ containing 20μg/mL RNase. (The solution should be cloudy with no obvious clumps.)

4. Add 200μL of freshly prepared Solution Ⅱ, mix gently (no vortex), and incubate 3~5min on ice (solution should be clear). Do not incubate for more than 5min.

5. Add 150μL of Solution Ⅲ, mix gently (no vortex), and incubate 5min on ice.

6. Centrifuge 15min at full speed in table-top centrifuge; pour off supernatant into 1.5mL Ep tube.

7. Extract with an equal volume of phenol equilibrated with 0.1mol/L Tris-Cl (pH8.0). Centrifuge for 5min at full speed (~12,000r/min). Transfer aqueous (upper) phase to new Ep tube.

8. Extract the samples with an equal volume of chloroform/isoamyl alcohol (24:1). Mix gently by hand to avoid shearing chromosomal DNA. Transfer aqueous (upper) phase to new Ep tube.

9. Add 0.1 volume of 3mol/L NaAc (pH5.2) and add 2.5 volumes of ice-cold anhydrous ethanol, mix gently and incubate at $-20°C$ for 15min to precipitate the DNA. Centrifuge 12,000r/min ×10min and pour off the supernatant.

10. Wash pellet with 1mL ice-cold 70% ethanol; centrifuge 5min in microfuge. Remove as much of the 70% ethanol as possible. Store the pellet of DNA in an open tube at the clean bench until ethanol evaporated.

11. Dissolve pellet in 50μL TE buffer, usually heat the DNA sample to 68°C for 10min to inactivate any contaminating DNase. Then store DNA at $-20°C$.

12. A small amount of plasmid DNA (10~15μL) was taken for identification by UV spectrophotometry and 1% agarose electrophoresis, respectively. (refer to Experiment 13 Part II.)

Notes

1. Solution I may be prepared as a 10×stock solution and stored $-20°C$ in small aliquots for later use. After the Solution I is added to the bacterial precipitation, make sure that the cells are thoroughly suspended by vortex.

2. Once the Solution II (SDS/ NaOH mixture) has been added to the bacterial suspension, it is imperative that the suspension be treated gently; chromosomal DNA are long and fragile, and vigorous treatment will readily damage the DNA molecules released from the bacteria.

Section III Integrative Experiments

Experiment title

Date
Observations and results

Discussion

    1. SDS-alkaline lysis method exploits the difference in _____ characteristics of plasmid DNA and chromosomal DNA fragments.

    2. The role of Solution II is to _____ .

    3. Adding Chloroform and Phenol in this method is to _____ .

    4. Will plasmid and chromosomal DNA be denatured under alkaline conditions (such as pH11~12)?

    5. What is the role of high concentration of potassium acetate and SDS?

Teacher's remarks
Signature
Date

Experiment 15　Isolation and Identification of Total RNA in Tissues/Cells

Part I　Isolation of Total RNA by TRIzol Reagent

Principle

TRIzol reagent is a ready-to-use reagent for the isolation of total RNA from cells and tissues, which includes a mono-phasic solution of phenol and guanidine isothiocyanate. During sample homogenization or lysis, TRIzol reagent maintains the integrity of the RNA, while disrupting cells and dissolving cell components. Addition of chloroform followed by centrifugation separates the solution into an aqueous phase and an organic phase. RNA remains exclusively in the aqueous phase. After transfer of the aqueous phase, the RNA is recovered by precipitation with isopropanol. After removal of the aqueous phase, the DNA and proteins in the sample can be recovered by sequential precipitation. Precipitation with ethanol yields DNA from the interphase, and an additional precipitation with isopropanol yields proteins from the organic phase.

Reagents

1. TRIzol reagent (purchase from a commercial company).
2. Chloroform.
3. Isopropanol.
4. 75% ethanol (in DEPC-treated ddH_2O).
5. RNase-free ddH_2O or 0.5% SDS solution.

To prepare RNase-free water, draw water into RNase-free glass bottles. Add diethylpyrocarbonate (DEPC) to 0.01% and leave overnight, then autoclave. The SDS solution must be prepared using DEPC-treated, autoclaved ddH_2O.

Procedure

1. Homogenization.

(1) Tissues: use a homogenizer to homogenize tissue samples in 1mL TRIzol Reagent per 50~100mg of tissue. The sample volume should not exceed 10% of the volume of TRIzol Reagent.

(2) Cells: for cells grown in monolayer, lyse directly in a culture dish by adding 0.5mL TRIzol Reagent to a 3.5cm diameter dish (1mL per 10cm^2 dish), and passing the cell lysate several times through a pipette. For cells grown in suspension, pellet cells by centrifugation and lyse cells by adding TRIzol reagent (1mL of the reagent per 5~10×10^6 of animal or yeast cells, or per 1×10^7 bacteria). An insufficient amount of TRIzol reagent may result in contamination of the isolated RNA with DNA.

2. Phase separation.

Incubate the homogenized samples for 5min at 15~30℃ to permit the complete dissociation of nucleoprotein complexes. Add 0.2mL of chloroform per 1mL of TRIzol reagent, cap sample tubes securely. Shake tubes vigorously by hand for 15s and incubate them at 15~30℃ for 2~3min. Centrifuge the samples at 12,000r/min for 15min at 2~8℃. Following centrifugation, the mixture separates into the lower phenol-chloroform phase, interphase and a colorless upper aqueous phase. RNA remains exclusively in the aqueous phase. The volume of the aqueous phase is about 60% of the volume of TRIzol reagent used for homogenization.

3. RNA precipitation.

Transfer the aqueous phase to a fresh Ep tube, and save the organic phase if isolation of DNA or protein is desired. Precipitate the RNA from the aqueous phase by mixing with isopropanol (use 0.5mL of isopropanol per 1mL of TRIzol reagent). Incubate samples at 15~30℃ for 10min and centrifuge at 12,000r/min for 10min at 2~8℃. The RNA precipitate, often invisible before centrifugation, forms a gel-like pellet on the side and bottom of the tube.

4. RNA wash

Remove the supernatant. Wash the RNA pellet once with 75% ethanol, adding at least 1mL of 75% ethanol, mix the sample by vortex and centrifuge at 10,000r/min for 5min at 2~8℃.

5. Redissolving RNA

Dry the RNA pellet for 5~10min in air. It is important not to let the RNA pellet dry completely as this will greatly decrease its solubility. Dissolve RNA in RNase-free water or 0.5% SDS solution by passing the solution a few times through a pipette tip, and incubating for 10min at 55~60℃. (Avoid SDS when RNA will be used in subsequent enzymatic reactions.)

Notes

1. When working with TRIzol reagent use gloves and eye protection (shield, safety goggles). Avoid contact with skin or clothing. Use in a chemical fume hood. Avoid breathing vapor.

2. Isolation of RNA from small quantities of tissue (1~10mg) or cells (10^2~10^5) samples: add 800μL of TRIzol to the tissue or cells. Add 200μg glycogen directly to the TRIzol (final concentration is 250μg/mL). To reduce viscosity, shear the genomic DNA with 2 passes through a 26 guage needle prior to chloroform addition. The glycogen remains in the aqueous phase and is co-precipitated with the RNA. It does not inhibit first-strand synthesis at concentrations up to 4mg/mL and does not inhibit PCR.

3. After homogenization and before addition of chloroform, samples can be stored at -60℃ to -70℃ for at least one month, the RNA precipitate (step 4, RNA wash) can be stored in 75% ethanol in 4℃ for at least one week, or at least one year at -5℃ to -20℃.

Part II Identification of RNA by UV Spectrophotometry

Principle

The RNA concentration is determined by measuring A_{260} against blank and then calculating the concentration of the measured sample using according factors. The absorption of 1 A is equivalent to approximately 40μg/mL for RNA. In the case of pure samples, the ratio A_{260}/A_{280} should be approximately 2.0.

You can calculate the concentration of the RNA in your sample as the following:

$$\text{RNA concentration (μg/mL)} = A_{260} \times (\text{dilution factor}) \times 40$$

Reagents

1. Distilled H_2O (dH_2O).

2. RNA solution (unknown concentration).

Procedure

1. See Experiment 13. The only difference is the dilution step (step 3): put 4μL of RNA with unknown concentration into 1mL of water and mix.

2. Take A_{280} readings, divide A_{260} by the A_{280}, and this should give a figure around 2.0. If it is less than 1.9, the RNA is not pure and sample should be purified using phenol/chloroform.

Part III Identification of RNA by Formaldehyde Denaturing Agarose Gel Electrophoresis

Principle

A denaturing gel system is suggested because most RNA forms extensive secondary structure via intramolecular base pairing, and this prevents it from migrating strictly according to its size. The mobility of RNA (or DNA) will not be affected by their base composition and conformation in the presence of base pairing inhibitors (formaldehyde, urea or formamide), and the relationship between mobility and common logarithm of molecular weight will be inversely proportional. Denaturing agarose gel electrophoresis can be used for RNA fractionation, Northern blotting, measuring the length of RNA (or DNA), and the recovery of oligonucleotide.

Reagents

1. Formaldehyde.
2. Formamide: purchase and store in small aliquots under nitrogen at $-20°C$.
3. 5×MOPS buffer (0.1mol/L MOPS, pH7.0, 40mmol/L NaAc, 5mmol/L EDTA): Dissolve 20.6g MOPS in 800mL DEPC-treated 50mmol/L NaAc solution, add enough 2mol/L NaOH to adjust pH to 7.0, and then add 10mL 0.5mol/L EDTA (pH8.0). Bring final volume to 1L by adding ddH_2O. Store at 4°C in the dark after autoclave.
4. 10×formaldehyde gel-loading buffer (50% glycerol, 1mmol/L EDTA pH8.0, 0.25% bromphenol blue, 0.25% green xylene), treated with 0.1% DEPC water and autoclave.
5. Ethidium bromide (EB, 0.5μg/mL) or SYBR Green II.

Procedure

1. Set up the denaturation reaction. In sterile microfuge tubes mix:
RNA (up to 20μg): 4.0μL;
5×MOPS buffer: 2.0μL;
formaldehyde: 4.0μL;
formamide: 10.0μL.

2. Incubate the RNA solutions for 15min at 65°C. Chill the samples for 5~10min on ice, and then centrifuge them for 5s to deposit all of the fluid in the bottom of the tubes. Add 2μL of 10×formaldehyde gel-loading buffer to each sample and place the tubes on ice.

3. Install the denaturing agarose gel containing formaldehyde in a horizontal electrophoretic tank. Add sufficient 1×MOPS buffer to cover the gel to a depth of approx. 1mm. Run the gel for 5min at 5V/cm, and then load the RNA samples into the wells of the gel.

4. Run the gel submerged in 1×MOPS buffer at 4~5V/cm until the bromophenol blue has migrated at least 2~3cm into the gel (1~2h).

5. Immerse the gel in 0.5μg/mL of EB solution for 10min after electrophoresis, and then destaining with 10mol/L $MgCl_2$ for 5min. Visualize the RNAs in the gel using a UV detector.

Notes

1. Formaldehyde is toxic through skin contact and inhalation of vapors. Manipulations involving formaldehyde

should be done in a chemical fume hood.

2. Intact total RNA run on a denaturing gel will have sharp 28S and 18S rRNA bands (eukaryotic samples). The 28S rRNA band should be approximately twice as intense as the 18S rRNA band. This 2:1 ratio (28S:18S) is a good indication that the RNA is intact. Partially degraded RNA will have a smeared appearance, will lack the sharp rRNA bands, or will not exhibit a 2:1 ratio.

Experiment title

Date
Observations and results

Discussion

1. TRIzol Reagent includes a mono-phasic solution of _____ and _____.
2. RNA is high sensitive to RNA enzyme (RNase), which has very stable biological activity. RNase activity does not require _____, and not influenced by _____.
3. The ratio A_{260}/A_{280} is used to estimate the purity of nucleic acid. Pure RNA should give a value of approximately _____.
4. How to control the contaminating RNase for RNA isolation?
5. Why is the denaturing gel system suggested for RNA electrophoresis?
6. Please give some base pairing inhibitors.
7. After total RNA isolation, it is required to identify the integrity of RNA by denaturing gel electrophoresis. Can we see mRNA bands under UV light by staining with ethidium bromide?

Teacher's remarks
Signature
Date

Experiment 16 Polymerase Chain Reaction (PCR) and Reverse Transcription-PCR

Protocol I Polymerase Chain Reaction (PCR)

Principle

PCR is a rapid procedure for in vitro enzymatic amplification of a specific segment of DNA. The purpose of PCR is to make a huge number of copies of a gene. PCR is based on the DNA polymerization reaction. A primer and dNTPs are added along with DNA template and *Taq* DNA polymerase. In addition to using a primer that sits on the 5′ end of the gene and makes a new strand in that direction, a primer is made to the opposite strand to go in the other direction. The amplification reaction involves three steps: ① denaturation: template DNA is heated to denature the dsDNA molecules, making them single-stranded; ② annealing: the reaction mix is then cooled, allowing primers to anneal to complementary sequences on opposite strands of the template DNA (by hydrogen bonding between complementary bases: A-T, G-C) flanking the DNA segment to be amplified; ③ extension: the reaction is then brought to an intermediate temperature, and using free deoxyribonucleotides added to the reaction mixture, DNA polymerase extends these primers from their 3′ ends toward each other. This three-step process is repeated for 20~40 cycles, resulting in the production of millions of copies of the template DNA.

Reagents

1. Diluted DNA template.
2. dNTP mix solution (2mmol/L each dATP, dTTP, dCTP, dGTP) (pH7.5).
3. 10×PCR Buffer mix (100mmol/L Tris-Cl, pH8.3, 500mmol/L KCl, 1.5mmol/L $MgCl_2$).
4. Sense primer (50μmol/L).
5. Antisense primer (50μmol/L).
6. *Taq* DNA polymerase (2~5 U/μL).
7. Nuclease-free water.

Procedure

1. Primers design.

The selection of primers is very important to the efficiency of the reaction. Usually the primers are custom synthesized based on the sequence of the DNA that is being amplified. Two primers would have to be made for each of the inserts.

2. Take three 0.2mL thin-walled microcentrifuge tubes and mark them with blank, positive and sample. Add the following reagents into each of the tubes:

36.0μL	H_2O
5.0μL	10×buffer mix
1.0μL	each primers (50μM)
5.0μL	dNTPs
1.0μL	*Taq* DNA polymerase (2.5U)
1.0μL	diluted DNA template (0.05~1μg) only into sample tube
50.0μL	Total volume

(you could add the master mix buffer and *Taq* DNA polymerase, The master mix contains all of the components necessary to make new strands of DNA in the PCR process. It could be obtained from a variety of commercial company.)

3. Place these three tubes in the thermocycler and start to run the PCR program for the amplification.

94℃	5min	DNA initial denaturation	
94℃	30s	denaturation	
58℃	40s	annealing	25 cycles
72℃	1min	extension	
72℃	5min	(for final extension)	
4℃	forever	(for storage)	

4. Analyze a 10μL aliquot of PCR products by running on a 1%~2% agarose gel electrophoresis.

Notes

1. The essential criteria for any DNA sample are that it contains at least one intact DNA strand encompassing the region to be amplified and that any impurities are sufficiently diluted so as not to inhibit the polymerization step of the PCR reaction. Usually a 1 : 5 dilution of the sample with water is sufficient to dilute out any impurities.

2. PCR involves preparation of the sample, the primers and PCR mixture system, followed by detection and analysis of the reaction products.

3. After the first few cycles, most of the product DNA strands made are the same length as the distance between the primers. After PCR, aliquots of the mixture typically are loaded onto an agarose gel and electrophoresed to detect amplified product, note the extent to which you diluted the DNA in the reactions. Reaction only has about 1 picogram of DNA template. For this reason, on the agarose gel you should only see the amplified fragment.

Protocol II Reverse Transcription-PCR (RT-PCR)

Principle

RT-PCR is the most sensitive method for mRNA detection and quantitation. RT-PCR can be used to quantify mRNA levels from much smaller samples. Traditionally RT-PCR involves two steps: the RT reaction and PCR amplification. RNA is first reverse transcribed into cDNA using a reverse transcriptase, and the resulting cDNA is used as templates for subsequent PCR amplification using primers specific for one or more genes. RT-PCR can also be carried out as one-step RT-PCR in which all reaction components are mixed in one tube prior to starting the reactions. Although one-step RT-PCR offers simplicity and convenience and minimizes the possibility for contamination, the resulting cDNA cannot be repeated used as in two step RT-PCR. Reverse Transcription here is carried out with the SuperScript First-Strand Synthesis System for RT-PCR.

Reagents

1. Oligo(dT)$_{12-18}$(100μg/mL) in TE (pH8.0) or random primers.
2. dNTP mix solution (10mmol/L) containing all four dNTPs (pH7.5).
3. Reverse transcriptase (it can be supplied by company).
4. 5×First strand buffer: 250mmol/L Tris-Cl pH8.3, 375mmol/L KCl, 15mmol/L MgCl$_2$ (This buffer is supplied by company).

5. 0.1mol/L DTT.
6. 10×PCR buffer: 200mmol/L Tris-HCl pH8.4, 500mmol/L KCl.
7. 50mmol/L $MgCl_2$.
8. Sense primer (10μmol/L) for amplification of cDNA by PCR.
9. Antisense primer (10μmol/L) for amplification of cDNA by PCR.
10. Taq DNA polymerase (2~5U/μL).
11. Nuclease-free water.

Procedure

1. First strand cDNA synthesis: A 20μL reaction volume can be used for 1~5μg of total RNA or 50~500 ng of mRNA. Add the following components to a nuclease-free microfuge tube:

    1μL Oligo(dT)$_{12-18}$ or 50~250 ng of random primers;

    1~5μg total RNA;

    Nuclease-free water to 12μL.

    Heat mixture to 70℃ for 10min and quick chill on ice. Collect the contents of the tube by brief centrifugation and add:

    4μL 5×RT buffer;

    2μL 0.1mol/L DTT;

    1μL 10mmol/L dNTP mix.

2. Mix contents of the tube gently and incubate at 42℃ for 2min.
3. Add 1μL (200 units) of SuperScript II, mix by pipetting gently up and down. Incubate 50min at 42℃.
4. Inactivate the reaction by heating at 70℃ for 15min, and then chill on ice. Store the 1st strand cDNA at −20℃ until use for PCR.
5. Add the following to a PCR reaction tube for a final reaction volume of 50μL:

5μL	10×PCR buffer
1.5μL	50mmol/L $MgCl_2$
1μL	10mmol/L dNTP Mix
1μL	sense primer (10μmol/L)
1μL	antisense primer (10μmol/L)
1μL	Taq DNA polymerase (2~5 U/μl)
1μL	cDNA (from first strand reaction, preferably RNase H-treated)
40μL	autoclaved, distilled water
50μL	Total volume

6. Mix gently, start to run the PCR program

94℃	2min	DNA initial denaturation	
94℃	45 s	denaturation	
55℃	45 s	annealing	15~40 cycles
72℃	1min	extension	
72℃	10min	for final extension	
4℃	forever	for storage	

    Times and temperatures may need to be adapted to suit the particular reaction conditions.

7. Analyze 5μL of the PCR product by agarose gel electrophoresis. The rest is stored at −20℃.

Notes

1. The cDNA from first strand reaction can be used as a template for amplification in PCR. However, amplification of some PCR fragments (>1kb) may require the removal of RNA complementary to the cDNA. To remove RNA complementary to the cDNA, add 1μL (2 units) of *E. coli* RNase H and incubate 37℃ for 20min.

2. *Taq* DNA polymerase is the standard and appropriate enzyme for the amplification stage of most forms of RT-PCR. However, where elongation from 3′-mismatched primers is suspected, a thermostable DNA polymerase with 3′~5′ proofreading activity may be preferred.

3. Use only 10% of the first strand reaction for PCR. Adding larger amounts of the first strand reaction may not increase amplification and may result in decreased amounts of PCR product.

Section III Integrative Experiments

Experiment title

Date
Observations and results

Discussion

1. There are three major steps in a PCR, which include _____, _____, and _____.
2. The purpose of a PCR is _____.
3. The template of RT-PCR is from _____.
4. Traditionally RT-PCR involves two steps: _____ and _____.
5. What is the principle of PCR?

Teacher's remarks
Signature
Date

Chapter 11　Genetic Engineering

Genetic engineering means the deliberate modification of an organism's genetic information by directly changing its nucleic acid genome and it is accomplished by a collection of methods known as recombinant DNA technology. The process of genetic engineering has four main steps: isolation of the genes of interest, insertion of the genes into a transfer vector, transformation of cells of organism to be modified, screen the genetically modified organism from those that have not been successfully modified. Genetic engineering makes genetic information transfer freely among different species, even between prokaryotic and eukaryotic, plants and animals, etc. The promise of its application is great in medicine, agriculture, and industry.

11.1　The Process of Genetic Engineering

Cloning is the process of making an identical copy of something. The term also covers when organisms such as bacteria, insects or plants reproduce asexually. Although PCR technique can clone large quantities of DNA fragments *in vitro*, gene cloning more often refers to the procedure of isolating a defined DNA sequence and obtaining multiple copies of it *in vivo*. It is a process by which large quantities of a specific, desired gene or section of DNA may be copied once the desired DNA has been isolated. It is often used to get enough amount of gene for further analysis. Although gene cloning is frequently employed to amplify DNA fragments containing special target genes, it can also be used to amplify any DNA sequence such as promoters, non-coding sequences and randomly fragmented DNA. It is utilized in a wide array of biological experiments and practical applications such as large scale protein production.

The basic procedures of genetic engineering include: ① isolating the target gene and selecting a suitable vector; ② ligation of the target gene and vector; ③ transformation of the recombinant DNA into host cells; ④ screening the positive host cells containing recombinant DNA; ⑤ expression and identification of the recombinant DNA product (Chapter 2).

11.1.1　Fragmentation/Isolation

Initially, the gene of interest which to be inserted into the organism needs to be identified and isolated to provide a DNA segment of suitable size. DNA fragments can be obtained from cDNA or genome DNA libraries. One can extract the genome DNA or total RNA from an organism, then isolate the fragment of interest by cleaving the DNA or cDNAs into fragments. Alternatively, the desired DNA fragment also can be directly synthesized by PCR and DNA synthesizing techniques. If necessary, i.e., for insertion of eukaryotic genomic DNA into prokaryotes, further modification may be carried out such as removal of introns or ligating with prokaryotic promoters. The target gene or DNA that is desired then is treated with restriction enzymes for further cloning with suitable vectors.

11.1.2　Ligation/Insertion

Subsequently, a ligation procedure is used once the target gene is isolated. Joining linear DNA fragments together with covalent bonds is called ligation. The vector which is frequently circular needs to be linearized before the ligation. It is treated with the same restriction endonuclease as the gene of interest to produce identical sticky ends or be supplemented to special linkers at the ends. After that, both fragments are incubated together, the recombinant DNA is formed with the help of DNA ligase.

11.1.3 Transformation/Transfection

Following ligation, the vector with the insert of target gene is transformed into host cells to replicate. Transformation includes chemical sensitization of cells, microinjection, electroporation, bacterial transformation or transfection/ infection using virus vectors, etc.

11.1.4 Screening/Selection

The transformed cells can be cultured and screened. As the aforementioned procedures are of particularly low efficiency, there is a need to identify the cells that have been successfully transformed with the recombinant vector containing the desired DNA sequence. Cloning vectors usually include selectable antibiotic resistance markers, which allow only cells in which the vector has been transformed, to grow on agar dishes with the proper antibiotics. Another screening method called α-complementary is based on color selection markers of plasmid using colorless compound X-gal medium. To confirm the correctness of recombination gene, PCR, restriction endonuclease analysis and/or DNA sequencing should be performed.

11.2 Expression of Foreign Genes and Purification of Recombinant Proteins

11.2.1 Expression of foreign genes in host cells

Currently, there are several well-developed expression vectors readily available from commercial sources. In general, the expression systems are mainly divided into two categories: prokaryotic expression system and eukaryotic expression system. The prokaryotic expression system is mainly based on *E. coli*, and it is also the most widely used expression system at present. Eukaryotic expression system includes mammalian cell expression, yeast expression system and insect expression system. *E. coli* expression system is simple to operate with high amount of protein expression. However, due to the lack of modification system, eukaryotic proteins are easy to form inclusion bodies, which can not guarantee the protein activity. The mammalian cell expression system can guarantee the activity of expressed proteins and carry out complex modification of proteins (such as complex glycosylation or acetylation), but because of the need to cultivate cells, the cost is high, and the screening of cell lines is difficult. In yeast expression system, protein expression is high, but only simple modified proteins can be expressed.

11.2.2 Purification of recombinant proteins

In general, purification of recombinant protein is relatively easier than normal extraction and purification of proteins from tissues as the recombinant protein is expressed at high level in host cells. The strategy of purification for recombinant protein depends not only on the physicochemical properties of proteins but also the expression manner of the foreign gene, e.g., insoluble inclusion bodies, soluble recombinant proteins, or fusion proteins.

11.2.2.1 Purification of recombinant proteins from insoluble inclusion bodies

High-level expression of many proteins in *E. coli* might result in insoluble inclusion bodies that are failed folding intermediates. However, the formation of inclusion bodies is not necessarily undesirable, and it even has several advantages in some cases. At first, inclusion bodies usually represent the highest yielding fraction of expressed protein. More importantly, inclusion bodies are easy to isolate to high purity by simple centrifugation which will tremendously facilitate the purification of the target protein. In addition, the expressed proteins in inclusion bodies are generally protected from proteolysis.

If the application is to prepare antibodies, inclusion bodies can be used directly as an antigen. However, if the purpose of recombinant expression is to produce functional proteins, the isolated inclusion bodies should be solubilized and refolded by a variety of well-developed methods. Solubilization and refolding methods usually involve the use of chaotropic agents, co-solvents or detergents. For example, once purified, inclusion bodies can be solubilized by denaturation with guanidine hydrochloride or urea.

11.2.2.2 Purification of soluble recombinant proteins

If the cloned gene is expressed as soluble form, it may be excreted or located in the periplasm, in the membrane fraction, or most commonly in the cytoplasm. As for such soluble proteins, the conventional protein purification strategies should be carefully followed. In general, the purification procedures are divided into two main stages: ① based on the localization of the recombinant proteins, different methods such as centrifugation and filtration are used to obtain an extract containing the desired recombinant protein in soluble form; ② multiple conventional purification methods including salt fractionation, chromatography, and electrophoresis may be carried out to get the purified protein. Moreover, for therapeutic proteins, more strict steps must be taken to remove minor contaminants.

11.2.2.3 Purification of fusion proteins

In some cases, the foreign gene is expressed as fusion protein to achieve efficient translation and improved folding, and avoid proteolysis. In this approach, the foreign gene is introduced into an expression vector 5' or 3' to a carrier sequence which codes for a carrier peptide or protein (fusion partner). The fusion partner can be as small as a short peptide such as FLAG and HA, and can also encode an entire protein such as glutathione-S-transferase (GST). Most of the common fusion partners serve as affinity tags, and these make the isolation of the expressed protein relatively easy. Proteins can often be purified to >90% by affinity chromatography without knowing their physicochemical characteristics in details. After purification, the fusion partner can be removed by specific proteolytic or chemical cleavage. Detection of fusion proteins is also a simple matter, since antibodies and colorimetric substrates are available for several of the commonly used fusion partners. Table 11.1 summarizes some of the characteristics of the most widely used fusion partners.

Table 11.1 **Commonly used fusion partners**

Tag	Tag Size	Purification	Detection	Cleavage
FLAG	1kU	Anti-FLAG resin	Anti-FLAG Ab	Enterokinase
6×His	1kU	NTA-agarose	Anti-6×His Ab	Enterokinase, Thrombin
Myc	1.2kU	Anti-Myc Ab resin	Anti-Myc Ab	Not available
Glutathione Stransferase, GST	26.5kU	Glutathione sepharose/agarose	Anti-GST Ab, CDNB substrate	Thrombin, Factor Xa

Ab: antibody.

Experiment 17 DNA Cloning

DNA cloning involves separating a specific gene or DNA segment from its larger chromosome and attaching it to a small molecule of vector DNA. The basic process of DNA cloning includes: isolation of target gene, selection and construction of vectors, digestion and ligation of target DNA and vector, transformation of target gene into receptor cell, screening for recombinant plasmids, and expression of cloned gene, etc.

Part I Target DNA Preparation

See Experiment 13, Experiment 14.

Part II DNA Digestion by Restriction Endonucleases

Principle

Restriction endonuclease(RE) is able to identify specific DNA sequences and cuts the dsDNA within or near identified sites. There are two kinds of cut patterns: sticky ends and blunt ends. Only type II of restriction endonucleases are used for DNA cloning. According to different purposes, two digestion systems may be used such as small amount of enzyme reaction or large number. The former was always used to identify gene fragment, however, the latter was always used to prepare gene fragment.

Reagents

1. ddH$_2$O (DNase-free).
2. Restriction endonuclease (5~10U/μL, purchase from company).
3. 10×Reaction buffer (the manufacturers normally provide the optimum buffer).
4. DNA (plasmid DNA or genomic DNA).

Procedure

1. Adding the following solutions to a new Ep tube (sterile):
   7μL ddH$_2$O;
   2μL 10×Reaction buffer;
   1μL Restriction endonuclease (10U);
   10μL plasmid DNA (1μg).
2. Incubate at 37℃ for 1~2h (usually less than 7h).
3. The digested results are identified by rapid agarose gel electrophoresis analysis, and then decide whether to terminate the reaction.
4. Reaction can be terminated in three ways.
   (1) EDTA can be added into a final concentration of 10mmol/L, through chelating Mg^{2+} to terminate reaction, or 0.1% SDS is added to denature endonuclease and terminate the reaction.
   (2) The digested DNA solution is incubated at 65℃ for 20min, but some of enzymes cannot be completely inactivated by heating.
   (3) The digested DNA is extracted by phenol/chloroform, and then precipitated by ethanol. This is the most effective method.

Notes

1. The ddH$_2$O is for variable volume. The total volume of 20μL contains 0.2~1μg of DNA in a typical reaction. If the digested DNA is more than 1μg, the total volume can be enlarged according to the ratio of the standard system.

2. A newly sterilized tube must be used every time you take enzyme. The plasmid DNA is finally added to prevent cross-contamination of reagents.

3. Most of the restriction endonucleases are stored in 50% glycerol buffer at -20℃. If glycerol concentration is more than 5% in hydrolysis reaction, the activity of restriction endonucleases will be inhibited. So, it should be accurately diluted 10 times or more.

4. The excessive endonuclease (2~5 times) can shorten time of reaction and obtain complete digesting effect, but it will lead to the decreasing of identification to the specific sequence.

5. In order to save restriction endonucleases, you can appropriately increase enzymatic hydrolysis time, but not too long, otherwise it will produce other enzymatic hydrolysis activities.

Part III Identification of target DNA fragment by Gel Electrophoresis and Extraction of DNA fragment

Principle

After digested by REs, the DNA fragments were separated by gel electrophoresis. The target DNA fragment and linear vector can be recycled from gel by phenol extraction or DNA adsorption kit. Crushing gel, immersing in eluate, and centrifugation is one of methods for recovering DNA from gels. Obtained DNA usually does not contain enzyme inhibitors and other impurities poisonous for transfection or microinjection. The recovery rate is 30% to 90% depending on the size of DNA fragment. This method is widely used for isolating single, double-stranded DNA and oligonucleotides from gels, which are suitable as hybridization probe, templates for chemical and enzymatic sequencing. Both agarose gel and non-denaturing polyacrylamide gel can be used for the separation and purification of fragment DNA. Fragments from 200bp to 10kb the agarose purification is ideal. For smaller fragments (20~400bp) the polyacrylamide purification is preferred.

Reagents

1. Crush and Soak Solution (500mmol/L NH$_4$Ac, 0.1% SDS, 0.1mmol/L EDTA).

Add 3.3g NH$_4$Ac, 0.1g SDS, 20mL 500mmol/L EDTA into ddH$_2$O and bring up to 100mL, store at room temperature.

2. 3mol/L NaAc, pH5.2.

24.6g anhydrous sodium acetate, adjust pH to 5.2 with acetic acid and bring up to 100mL with ddH$_2$O, store at room temperature.

3. DMCS treated glass wool (Alltech Assoc. Inc. #4037, 50g).

4. Blue tips with melted tips to serve as pestle for crushing acrylamide.

5. Gel:

(1) Agarose gel with appropriate concentration for fragment DNA.

(2) Reagents for PAGE:

   Acrylamide: bisacrylamide stock solution (29:1);

   10% ammonium persulfate;

   6×gel-loading buffer;

5×TBE electrophoresis buffer;
10% TEMED.

Procedure

1. Agarose gels.

(1) After digestion, the samples were loaded into agarose gel, and separated by electrophoresis.

(2) Prepare spin columns by cutting off the cap of a 0.5mL Eppendorf (Ep) tube and forming a hole in the bottom with a hot 18Ga needle. Fill this "mini-column" with a small ball of DMCS treated glass wool and pack down with a pipet tip.

(3) Cut out the desired band from an agarose gel and place in a spin column inside a 1.5mL Ep tube with the top cut off.

(4) Centrifuge at 6,000r/min for 10min.

(5) Phenol/chloroform extracts the flow through and Ethanol precipitates with glycogen or tRNA and 10% V/V of 3mol/L NaAc.

(6) Wash and dry, resuspend in 20μL TE.

2. PAGE.

(1) Assemble the glass plates with spacers. Taking into account the size of the glass plates and the thickness of the spacers, calculate the volume of gel required.

(2) Prepare the non-denaturing polyacrylamide gel solution with the desired polyacrylamide percentage according to the table below, which gives the amount of each component required to make 10mL.

Recipes for polyacrylamide gels of indicated concentrations in 1×TBE

Gel concentration Reagents (mL)	3.5%	5.0%	8.0%	12.0%	20.0%
30% Acrylamide stock solution	1.16	1.66	2.66	4.0	6.66
H_2O	6.7	6.2	5.2	3.86	1.2
5×TBE	2.0	2.0	2.0	2.0	2.0
10% Ammonium persulfate	0.1	0.1	0.1	0.1	0.1

(3) Add 50μL of 10% TEMED for each 10mL of acrylamide: bis solution, mix the solution by gentle swirling, and then pour the gel solution into the space. (Wear gloves. Work quickly to complete the gel before the acrylamide polymerizes.)

(4) Immediately insert the appropriate comb into the gel, being careful not to allow air bubbles to become trapped under the teeth. Make sure that no acrylamide solution is leaking from the gel mold. Allow the acrylamide to polymerize for 30~60min at room temperature.

(5) When ready to proceed with electrophoresis, squirt 1×TBE buffer around and on top of the comb and carefully pull the comb from the polymerized gel. Use a syringe to rinse out the wells with 1×TBE.

(6) Fill the reservoirs of the electrophoresis tank with electrophoresis buffer. Use a Pasteur pipette or syringe needle to remove any air bubbles trapped beneath the bottom of the gel.

(7) Use a syringe to flush out the wells once more with 1×TBE. Mix the DNA samples with the appropriate amount of 6×gel-loading buffer. Load the mixture into the wells.

(8) Connect the electrodes to a power pack (positive electrode connected to the bottom reservoir), turn on the power, and begin the electrophoresis run.

(9) Run the gel until the marker dyes have migrated the desired distance. Turn off the electric power, disconnect the leads, and discard the electrophoresis buffer from the reservoirs.

(10) Detach the glass plates. Lay the glass plates on the bench (siliconized plate uppermost). Use a spacer or plastic wedge to lift a corner of the upper glass plate. Check that the gel remains attached to the lower plate. Pull the upper plate smoothly away. Remove the spacers.

(11) Carefully transfer the gel into a plastic dish and add EB staining solution to cover the gel. Set in a dark drawer for 30min. Visualize the DNA fragments on a long wave UV light box, and cut out the desired band. Crush the acrylamide with a p1,000 tip with a melted end to resemble a pestle for the eppendorf "mortar".

(12) Add 1mL crush and soak solution and incubate overnight at 37°C.

(13) Spin in the microfuge for 10min at 14,000r/min. Remove as much liquid as possible and add another 0.5mL of crush and soak solution.

(14) Repeat the spin and pool the recovered supernatant.

(15) Add 0.1 volume of 3mol/L NaAc, 2.5 volumes of Ethanol and carrier (see above).

(16) Spin as usual, wash and dry. Resuspend in 20μL TE.

Notes

1. Polyacrylamide gels are poured and run in 0.5× or 1×TBE at low voltage (1~8V/cm) to prevent denaturation of small fragments of DNA. Other electrophoresis buffers such as 1×TAE can be used, the gel must be run more slowly in 1×TAE, which does not provide as much buffering capacity as TBE. For electrophoresis runs greater than 8h, we recommend that 1×TBE buffer be used because adequate buffering capacity is available.

2. It is important to use the same batch of electrophoresis buffer in both of the reservoirs and in the gel. Small differences in ionic strength or pH produce buffer fronts that can greatly distort the migration of DNA.

3. Usually, approx. 20~100μL of DNA sample is loaded per well depending on the size of the slot. Do not attempt to expel all of samples from the loading device, as this almost always produces air bubbles that blow the sample out of the well. In many cases, the same device can be used to load many samples, provided it is thoroughly washed between each loading. However, it is important not to take too long to complete loading the gel; otherwise, the samples will diffuse from the wells.

4. Non-denaturing polyacrylamide gels are usually run at voltages between 1V/cm and 8V/cm. If electrophoresis is carried out at a higher voltage, differential heating in the center of the gel may cause bowing of the DNA bands or even melting of the strands of small DNA fragments. Therefore, with higher voltages, gel boxes that contain a metal plate or extended buffer chamber should be used to distribute the heat evenly. Alternatively, use a gel-temperature-monitoring strip.

Part IV Sticky End Ligation of Recombinant DNA

Protocol I Sticky End Directional Ligation of Recombinant DNA

Principle

DNA ligase can join the 3′ hydroxyl and adjacent 5′ phosphate group of DNA molecules to form phosphodiester bond. In this way, the digested target gene and the vector are connected to produce a new molecule called recombinant DNA. The enzyme used to ligate DNA fragments is T_4 DNA ligase, which originates from T_4 bacteriophage. This enzyme will ligate DNA fragments having overhanging, sticky ends that are annealed together or fragments with blunt ends. Here using the target gene and the vector containing *Bam*H I and *Eco*R I sites as an example, the digested target gene and vector can form the same sticky ends. The target gene and the vector are separately digested with *Bam*H I and *Eco*R I. And then recover these two digested DNA fragments by agarose gel electrophoresis. Afterwards, join them together by using T$_4$ DNA ligase. The ligation of the double enzyme digestion is also called directional cloning, because the insertion of the target gene into the vector has only one direction. The basic process is indicated as Fig. 11-1.

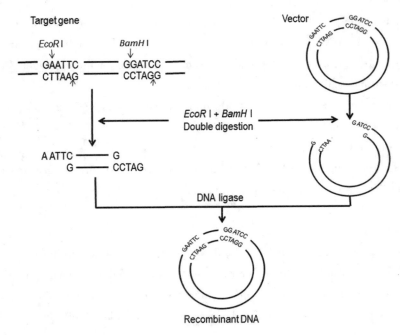

Fig. 11-1　Sticky end ligation

Reagents

1. ddH$_2$O (DNase-free).
2. 10×RE Reaction buffer (the manufacturers normally provide the optimum buffer).
3. DNA (plasmid DNA or genomic DNA).
4. *Bam*H I (5U/μl).
5. *Eco*R I (5U/μl).
6. T$_4$ DNA ligase (2U/μl).
7. 10×ligation buffer (the manufacturers normally provide the optimum buffer).
8. DNA (plasmid DNA or genomic DNA).

Procedure

1. Preparing the target gene fragment and vector, taking each 1~3μg, and then establishing the following reactions:

DNA: 1~3μg;
10×buffer: 2μL;
*Bam*H I (5U/μL): 1μL;
*Eco*R I (5U/μL): 1μL.

ddH$_2$O is added to total volume of 20μL. The reaction tube is incubated at 37℃ for 12~16h or overnight and then the gene fragments and linear vector are recycled by agarose gel electrophoresis.

2. Establishing ligation reaction in 1.5mL Ep tubes:

Target gene fragment: 0.4μg;
DNA vector: 0.1μg;
10×ligation buffer: 2μL;
T$_4$ DNA ligase (2U/μL): 1μL.

The ddH$_2$O is added to total volume of 20μL. The reaction tube is incubated at 16℃ for 12~16h or overnight. By this method, the target gene is inserted into the vector in the correct direction.

3. Recombinant plasmid is transformed into host cells (such as *E. coli*), and then cloned, screened and identified.

Protocol II TA cloning

Principle

The target DNA can be directly linked to T-vector with T-A pairing by DNA ligase. Because the PCR products amplified by *Taq* DNA polymerase will be added a deoxyadenosine (dA) at the 3' end, while the linearized T-vector has a deoxythymidine (dT) at the 3' end. For example, PMD18-T vector is constructed by adding *EcoR* V recognition sequence in MCS of pUC18 vector. The T-Vector will be linearized with 3' protruding T ends of both sides after cleaved by *EcoR* V. The pMD18-T vector also maintains the selective markers, including an ampicillin resistant (*Amp^r*) gene and a β-galactosidase (*lac* Z) gene, which could be used for Amp plate screening or blue-white screening. See Fig. 11-2.

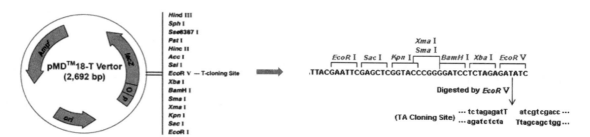

Fig. 11-2 The map of pMD18-T vector

Reagents

1. pMD18-T vector.
2. 10 × ligase buffer.
3. T4 DNA ligase (2U/μL).
4. Double distilled water (ddH$_2$O).
5. PCR product purification kit.

Procedure

1. Target gene obtained by PCR. (See Experiment 16.)
2. According to the instruction of purification kit, add the corresponding volume of Binding Buffer to each PCR tube, mix sample well.
3. Insert a high pure filter tube into one collection tube. Transfer the mixture to the filter tube, centrifuge 1min at 10,000g.
4. Discard the flowthrough solution in the Collection Tube. Add 500μL wash buffer to the filter tube, centrifuge 1min at 10,000g. Discard the flowthrough solution, repeat the washing with 200μL wash buffer.

5. Discard the flowthrough solution, centrifuge filter tube 1min at 10,000g for emptying tube.

6. Put the Filter Tube to a clean 1.5mL Ep tube. Add 50μL Elution Buffer to dissolve the DNA for 5min at room temperature, centrifuge 1min at 10,000g, then the purified PCR product is in the Ep tube.

7. The target gene from PCR product is ligated to the T vector.

The following reaction was established in a sterile Ep tube:

PCR product (0.1~0.3pmol/L): 1μL

PMD18-T vector (50ng/μL): 1μL

10 × ligase buffer: 2μL

T_4 DNA ligase (2U/μL): 1μL

ddH_2O: 15μL

Mix well, put the reaction tube in water bath at 16℃ for 12~16h or overnight, and the target gene should be inserted into the vector. Then the reaction solution containing recombinant DNA is collected for transformation.

Notes

1. The ratio of the target gene and vector DNA molecule is normally (3~5) : 1 in DNA ligation reaction. If there are too many target DNA molecules, polymers will be easily produced. The amount of the target DNA can be calculated according to the following formula:

The target DNA (ng) = vector(ng) × target DNA fragment(kb) × (3~5) ÷ vector length(kb)

2. When the target DNA is cloned to the linear vector with the same sticky ends, it is necessary to remove the 5'-P of vector for controlling self-cycling of the plasmid.

3. The optimal incubation temperature for T_4 DNA ligase is 16℃. However, this ligase is active at a broad range of temperatures: ligations performed at 4℃ overnight or at room temperature for 30min to a couple of hours usually work well.

4. Both *Bam*H I and *Eco*R I have relatively severe star activity when the ratio of restriction enzyme to DNA is too high. So, you should not use excess amount of enzyme to cut on those vectors, otherwise, it might introduce the star activity to the reaction and those enzymes randomly cut somewhere.

5. Heat inactivates the ligase by placing tube in 65℃ water bath for 10min.

Part V Preparation of Competent Cells and Transformation of Plasmids

Principle

Transformation is to introduce a foreign plasmid into a bacterium and to use this bacterium to amplify the plasmid in order to make large quantities of it or express the target gene. A plasmid is a small circular piece of DNA that may contain ampicillin- or tetracycline-resistant genes. A plasmid containing resistance to antibiotics is used as a vector. The gene of interest is inserted into the vector plasmid and this newly constructed plasmid is then put into *E. coli* that is sensitive to antibiotic (such as ampicillin). The bacteria are then spread over a plate that contains ampicillin. Finally, the bacteria transformed with plasmids which contain ampicillin-resistant gene can grow in ampicillin plate.

During transformation, recombinant DNA molecules must be introduced into a host to carry out amplification. Since DNA is a very hydrophilic molecule, it won't normally pass through a host cell's membrane or it is lowly taken up by general host cells. Successful transformation is difficult. In order to make bacteria take up the plasmid, they must first be made "competent" to take up DNA. After the bacteria are treated by a high concentration of calcium chloride, its membrane permeability changed so that its uptaking capacity of DNA increases considerably. These cells are called the competent bacteria. The transformed bacteria are cultured in the medium with appropriate antibiotics. The bacteria containing transformant can grow to bacterial colonies.

Reagents

1. LB liquid medium:

pancreatic peptone: 10.0g;

yeast extract: 5.0g;

NaCl: 10.0g.

The substrate is completely dissolved in ddH$_2$O by using magnetic stirring, and then the pH is adjusted to 7.0 by using 5mol/L NaOH. The capacity is determined to 1L, and then autoclaved at 15lbf/in^2 for 20min.

2. LB plate: 1.5% agar powder is added into LB liquid medium and autoclaved at 15lbf/in^2 for 20min, and cooled to about 60℃. Then add ampicillin of 50~100μg/mL and pour the solution into culture plate. Put LB plate at 4℃ if not used immediately.

3. 0.1mol/L CaCl$_2$ solution: 1.1g of anhydrous CaCl$_2$ is dissolved in 60mL of ddH$_2$O, and then the capacity is determined to 100mL. The solution is autoclaved, and then stored at 4℃.

4. Ampicillin (Amp): 100mg/mL Amp was stored at -20℃, and diluted to 50~100μg/mL for application.

5. Single colony or frozen *E. coli*.

Procedure

1. Preparation of competent cells.

(1) A single colony (e.g., DH$_{5α}$) is transferred to the tube with 3mL of LB medium, and then shaking at 37℃ overnight. The next day, 1mL bacterium solution are incubated to a 500mL flask with 100mL LB medium, and then vigorously shaking at 37℃ (200~300r/min) for 2~3h. After A_{600} value reaches 0.3~0.4, the flask is immediately removed to ice bath for 10~15min.

(2) Sterile operation is required from this step. Bacteria are transferred to a 50mL pre-cooled sterile tube.

(3) Centrifuge (4,000g) for 10min at 4℃, then discard supernatant.

(4) Add pre-cooled 10mL CaCl$_2$(0.1mol/L) to suspend cells. The suspended cells are put in ice bath for 30min.

(5) Centrifuge (4,000g) for 10min at 4℃, then discard supernatant.

(6) Add pre-cooled 4mL CaCl$_2$(0.1mol/L), and gently resuspend cells.

(7) Immediately dispense every 0.2mL suspended cells into a 0.5mL autoclaved Ep tube and stored at 4℃ for 12~24h, which can increase competency.

2. To freeze competent cells: competent cells can be stored at -70℃ for future transformation. Aliquot the competent cells (200~400μL), add 30% glycerol, and the competent cells can be stored at -70℃ for several months.

3. Transformation:

(1) Add 100μL the competent cells into a sterile Ep tube.

(2) Add 50~100ng plasmid to the above solution, the volume should not be more than 5% of competent cells. The mixture is gently rotated, and then place on ice for 30min.

(3) Heat shock at 42℃ for 90s, and then cool the solution about 1~2min on ice.

(4) Add 400μL LB medium without antibiotics into each tube, and shake at 37℃ for 45~60min in a shaker (100~150r/min). So that bacteria will be reanimated, and express the resistance gene of plasmid.

(5) 100μL of bacterial solution is spread on LB plate containing Amp of 50~100μg/mL. Put plates at 37℃ for 15~20min. Then incubate plates upside down at 37℃ for 10~16h (not more than 20h). (This step is one of screening methods which using antibiotics resistance markers.)

Notes

1. The value of A_{600} should be controlled between 0.3 and 0.4.
2. The used plasmid should be of covalently closed-loop DNA.
3. $CaCl_2$ should be of high quality.
4. Prevent contamination from other bacteria and exogenous DNA.
5. A positive control containing competent cells and plasmid DNA should be established in the experiment to estimate transformation efficiency. A negative control containing only competent cells should be established to eliminate potential contamination. If cloning can grow in a negative control, the possible causes are: ① the competent cells are contaminated by the resistant strains; ② the selective plates failed; ③ the selective plates are contaminated by the resistant strains.
6. No colonies obtained after transformation could be due to inactive competent cells or unsuccessful ligation. Therefore you can check:

(1) Competent cells activity. Perform a test transformation using a known amount of standard supercoiled plasmid DNA as a positive control.

(2) Check the ligase activity. If you have enough ligation products left, analyze linearized vector DNA and ligated product on an agarose gel.

(3) Optimize transformation condition. The added ligation products should be less than 5% of the transformation volume. Excess DNA may inhibit the transformation.

Part VI Identification of Recombinant Plasmid

After recombinant plasmids are transformed into *E. coli*, the transformed bacteria need to be screened and identified for selecting recombinant plasmids in which the target gene is inserted according to correct direction. The following are four common methods: ① α-complementary; ② inserting inactivation; ③ *in situ* hybridization; ④ restriction endonuclease analysis. The first three screening methods only identify if a gene fragment has been inserted in recombinant plasmids, restriction endonuclease analysis can identify inserting direction.

Protocol I α-complementary

Principle

Vector contains the regulatory sequences of β-galactosidase gene (*LacZ*) and the encoded sequences of 146 amino acids at N-terminus, such as pUC series and pGEM-3Z. Multiple cloning sites (MCS) are inserted in the encoding region. These vectors can be applied to transform the host cells that can encode C-terminal of β-galactosidase. Neither the encoded fragments by plasmid nor the host cells have enzyme activity, but they can be integrated to recover an enzyme activity. Integrated *LacZ* can catalyze X-gal to form blue colonies. This phenomenon is called as α-complementary. When the foreign DNA is inserted into multiple cloning sites of plasmid, it will result in inactivation of N-terminal fragment. Therefore, white colonies must contain the recombinant plasmid.

Reagents

1. X-gal (20mg/mL): 20mg X-gal is dissolved in 1mL dimethyl formamide, and stored at $-20^\circ C$ in the dark.
2. IPTG (isopropyl β-D-thiogalactoside) (1mol/L): 2.38g IPTG is dissolved in 10mL ddH_2O, filtered by 0.22μm membrane, and then stored aliquot at $-20^\circ C$.

Procedure

1. Prepare agar plates which contain the corresponding antibiotics (such as Ampicillin).
2. 40μL X-gal and 4μL IPTG are added onto a plate surface, and evenly spread by using sterile glass stick. Then incubate at 37℃ for 1h to absorb the solution.
3. 200μL transformed bacteria solution is spread on the plate surface. Put plates at 37℃ for 15~20min, and then incubate plates upside down at 37℃ for 12~16h.
4. When stop culturing, incubate the plates at 4℃ for 1~4h. Blue colonies are negative, while white colonies contain the recombinant plasmids.
5. Pick up white colonies into 5mL LB medium with corresponding antibiotics such as Amp, and then shake at 37℃ for 8~12h (200~300r/min). The plasmids are extracted to further identify by restriction endonuclease analysis.

Protocol II Isolation of plasmids and restriction endonuclease analysis

Procedure

1. Pick up colonies to LB medium with Amp, and then culture them at 37℃ overnight.
2. Prepare plasmid DNA by alkaline lysis method (see Experiment 16).
3. Restriction endonuclease analysis (see Part II).
4. Identify the DNA fragment by non-denaturing polyacrylamide gel electrophoresis (see Part III)

Protocol III PCR analysis of target DNA fragment

The target DNA fragment in recombinant plasmids can be amplified by PCR with specific primer of the target DNA, and then be detected by gel electrophoresis. The protocol of PCR refers to see Experiment 16.

Section III Integrative Experiments

Experiment title

Date
Observations and results

Discussion

1. What is the competent cells?
2. Which characteristics are required for a typical plasmid (vectors) used for molecular biology?
3. How many steps are involved in DNA Cloning?
4. There are four common methods for the identification of recombinant plasmid, which include _____, _____, _____, and _____.

Teacher's remarks
Signature
Date

Experiment 18 Expression of Exogenous Gene in *E. coli* and Purification of His-tag Proteins

Part I Identification of Exogenous Gene Expression by SDS-PAGE

Principle

One of the basic methods used to raise the expression level of foreign gene is to separate the growth of host bacteria and the expression of foreign gene into two phases, which can reduce the burden on host bacteria. The temperature-induced and drug-induced agents are commonly used to improve the expression level of foreign gene. For example, expression vectors with Lac or Tac promoter (such as vector pET-His) can be induced by IPTG. Exogenous protein levels can be detected by SDS-PAGE and Western blotting.

The lysis of bacteria is a key step to extract intracellular products. The application of lysozyme and ultrasonication is popular in laboratory studies. Lysozyme method is to damage the bacterial membrane using lysozyme. Ultrasonication is mechanical crushing, which use ultrasound of more than 15~20kHz to disrupt cells.

Reagents

1. LB medium (see Experiment 21).
2. $BL_{21}(DE_3)$ strains of *E. coli* competent cells.
3. Vector pET-His containing the target gene.
4. IPTG stock solution (1mol/L).
5. Lysis buffer:
   50mmol/L Tris-Cl (pH8.0);
   1mmol/L EDTA;
   100mmol/L NaCl;
6. PMSF: 50mmol/L.
7. Lysozyme: 10mg/mL.
8. Sodium deoxycholate (DOC).
9. *DNase* I: 1mg/mL.
10. The reagents for SDS-PAGE.

Procedure

1. The pET-His containing the target gene was transformed into $BL_{21}(DE_3)$ strains of *E. coli*, and the positive colonies were obtained by screening (see Experiment 21). Plasmid DNA was purified to carry out restriction endonuclease analysis and sequencing.

2. Pick up one positive colony into 3~5mL LB/Amp medium, and then shake at 37℃ overnight (200~300r/min). Next day, take 0.1mL bacteria to 10mL LB/Amp medium, and shake at 37℃ for 2~3h (200~300r/min). When the value of A_{600} reaches at 0.3~0.4, add IPTG to a final concentration of 1mmol/L, and keep on culturing for 3~5h. At the same time, the bacteria without IPTG-induced can be as control.

3. Take 10mL with or without IPTG-induced bacteria respectively, and then centrifuge 5,000r/min at 4℃ for 15min.

4. Discard supernatant, and resuspend the pellet by lysis buffer (3mL per gram wet weight of bacteria).

5. Add PMSF 8μL and lysozyme 80μL per gram bacteria, and stir using glass stick for 20min at 37℃. When

stirring, add DOC 4mg per gram bacteria.

6. When the solution becomes viscous, add 20μL *DNase* I per gram bacteria. Incubate the solution at room temperature until no longer sticky.

7. (Optional) Bacteria can also be broken further by a sonicator according to the ultrasonic functional parameters. When the supernatant become translucent from turbid, the sonicating finished.

8. Centrifuge 10,000g for 15min, and collect the supernatant and pellet respectively. Take a small volume of the supernatant and pellet, add equal volume of 2×SDS-PAGE loading buffer to the supernatant and 1×SDS-PAGE loading buffer to the pellet, pipetting up and down, and heat at 100℃ for 3~5min. Then the samples were identified by SDS-PAGE.

Notes

1. Different expressive plasmids have different promoters, so the inducing methods are not exactly the same. If the vectors have PL promoter, they can be induced by temperature. The bacteria were cultured at 30~32℃ for a few hours so that A_{600} reached at 0.4~0.6, then temperature was quickly raised to 42℃, and kept on culturing for 3~5h.

2. Ultrasonication should be controlled according to strain type, cell concentration, audiofrequency sonic energy, etc. If the cells were repeatedly frozen and thawed 3~4 times before ultrasonic treatment, they would be broken more easily.

3. If the target protein is mainly located in the supernatant, it is expressed as soluble form. As for such soluble proteins, the conventional protein purification strategies can be used. If the target protein is mainly located in the pellet, it means the formation of inclusion bodies. Next are the purification strategies for these insoluble proteins.

Part II Identification of target protein by Western blotting

Before be carried to the purification, the specific exogenous protein should be detected by Western blotting to ensure the expression in host bacteria, the protocol refer to Experiment 12.

Part III Isolation, Dissolving and Renaturation of Inclusion Body in *E. coli*

Step I Isolation of inclusion body

Principle

When exogenous proteins were highly expressed in bacteria, they were often gathered to form inclusion bodies in the cytoplasm. After bacteria were lysed, they were released, and then collected by centrifuging. Inclusion bodies are usually not soluble in water, but can be dissolved by adding protein-denaturing agent such as urea or guanidine hydrochloride. When the expressed proteins contain cysteine, disulfide bonds can be formed in the inclusion bodies. Under such circumstances, the expressed proteins can be completely reduced by denaturant, DTT or β-mercaptoethanol. With the decreasing of denaturant concentration, the expressed proteins can resume their natural configuration, which is called the refolding of proteins with inclusion body. In order to form the correct disulfide bond, the reduced/oxidized glutathione redox buffer was always used to refold. The correct refolding is the most critical and most complex problem in gene engineering.

Reagents

1. LB medium.
2. $BL_{21}(DE_3)$ strains of *E. coli*.

3. Vector pET-His containing the target gene.

4. 1mol/L IPTG stock solution.

5. Lysis buffer:

   50mmol/L Tris-Cl (pH8.0);

   1mmol/L EDTA;

   100mmol/L NaCl.

6. Solution I: 0.5% Triton X-100 and 10mmol/L EDTA (pH8.0) were dissolved in the lysis buffer.

7. Urea solution.

Protocol I Triton-X 100/EDTA treatment

Procedure

1. Take 10mL IPTG-induced bacteria, and then centrifuge 5,000r/min at 4℃ for 15min.

2. Discard supernatant, and resuspend the pellet by lysis buffer (3mL/g wet weight of bacteria).

3. Centrifuge 12,000g at 4℃ for 15min. Discard the supernatant.

4. Resuspend the pellet by adding 9.0 volumes of solution I, incubate at room temperature for 5min, and then centrifuge 12,000g at 4℃ for 15min.

5. Reserve the supernatant, and resuspend the pellet by 100μL ddH$_2$O.

6. Take a small volume of the supernatant and resuspended pellet, add equal volume of 2×SDS-PAGE loading buffer, and heat at 100℃ for 3~5min. The samples were identified by SDS-PAGE.

Protocol II Urea treatment

Procedure

1. Centrifuge cell lysate (see above) 12,000g at 4℃ for 15min. Discard the supernatant.

2. Resuspend the pellet by 1mL ddH$_2$O per gram becteria. Taking 100μL ddH$_2$O respectively with 4 Ep tubes, the remaining is reserved.

3. Centrifuge the solution 12,000g at 4℃ for 15min. Discard the supernatant.

4. Resuspend the pellet by 100μL urea solution [containing different concentrations of urea (0.5, 1, 2, 5mol/L) in 0.1mol/L Tris-Cl (pH8.5)].

5. Incubate at room temperature for 5min, and then centrifuge 12,000g at 4℃ for 15min.

6. Reserve the supernatant, and resuspend the pellet by 100μL ddH$_2$O.

7. The supernatant and resuspended pellets were detected by SDS-PAGE (see above).

Step II Dissolving and renaturation of inclusion body

Reagents

1. Buffer I:

   0.1mmol/L PMSF;

   8mol/L Urea;

   10mmol/L DTT (dissolved in lysis buffer).

2. Buffer II:

   50mmol/L KH$_2$PO4 (pH10.7);

   1mmol/L EDTA (pH8.0);

50mmol/L NaCl;

1mmol/L reduced glutathione (GSH);

1mmol/L oxidized glutathione (GSSG).

3. 1mol/L KOH.

4. 1mol/L HCl.

Procedure

1. Inclusion bodies (the pellet obtained from step 1) were dissolved by 100 μL buffer I, and put at room temperature for 1h.

2. Add 9-fold volumes of buffer II, put at room temperature for 30min, and then adjust pH to 10.7 with KOH.

3. Adjust pH to 8.0 with HCl, and put at room temperature for 30min at least. Centrifuge 12,000g at room temperature for 15min.

4. Reserve the supernatant, take a small volume, add 2×SDS-PAGE loading buffer, and then add 1×SDS-PAGE loading buffer to the pellet. Samples were detected by SDS-PAGE.

Notes

In order to enhance the efficiency of protein refolding, the followings should be noted in dissolving inclusion body: ① maximize the purity of inclusion body; ② fully dissolve inclusion bodies, and reduce proteins completely; ③ if necessary, add reductants; ④ the concentration of denaturants should be reduced little by little, and maintained at the proper concentration to inhibit protein aggregation and wrong intermolecular disulfide bond formation; ⑤ application of redox buffer; ⑥ some factors should be adjusted to the most optimal, such as protein concentration, pH, temperature, the type of salt, ionic strength, and so on.

Part IV Expression and Purification of His-tagged Fusion Protein

Principle

Affinity chromatography can be performed using a number of different protein tags, such as poly-hisitidine (His) tagged proteins. The histidine tag is very short (6~10 his residues) and should not alter the conformation of the tagged protein, nor should it be involved in artifactual interactions. The poly-his tag binds to a nickel (Ni) chelate resin for creation of the column.

The fusion proteins are always the insoluble inclusion bodies (IB), which can prevent from the hydrolysis of protease to the expressed proteins and facilitate to isolate the expressive products using centrifuge. 6mol/L Guanidine-HCl, 8mol/L Urea or other strong denaturants can be used to completely solubilize IB. Since under denaturating conditions the His tag is completely exposed, it will facilitate the binding to Ni columns. For most biochemical studies, proteins have to be renatured and refolded, and this can be done in the column itself before elution or after elution.

Reagents and material

1. Vector pET-His containing the target gene.

2. BL21 (DE3) strains of *E. coli* competent cells.

3. LB medium.

4. 1mol/L IPTG stock solution.

5. 10×PBS buffer.

6. Ni-NTA agarose beads.

7. Lysis buffer: 50mmol/L Na$_2$HPO$_4$ pH8.0, 0.3mol/L NaCl, 1mmol/L PMSF (or protease inhibitor cocktail for bacterial cells #P-8849 from Sigma) and strong denaturant as 6mol/L Guanidine-HCl or 6~8mol/L urea.

8. Equilibration buffer: 6~8mol/L urea, 50mmol/L Na$_2$HPO$_4$ pH8.0, 0.5mol/L NaCl.

9. Washing buffer: 6~8mol/L urea, 50mmol/L Na$_2$HPO$_4$ pH8.0, 0.5mol/L NaCl.

10. Elution buffer: 6~8mol/L urea, 20mmol/L Tris pH7.5, 100mmol/L NaCl, and appropriate imidazole concentrations.

11. The reagents for SDS-PAGE.

Procedure

1. The vector pET-His containing the target gene was transformed into competent *E. coli*. The transformant was screened in LB plate with Amp (see Experiment 21). Pick the positive colony into 2mL LB medium with Amp and incubate at 37℃ for 3~5h.

2. Add 1mol/L IPTG to a final concentration of 0.1mmol/L, and then culture at 37℃ for 3~5h to induce expression of the fusion protein.

3. Centrifuge the bacteria 10,000g at room temperature for 5s. Remove the supernatant. Resuspend the pellet by precooled PBS of 300μL.

4. Disrupt the bacteria by ultrasonic, centrifuge 10,000g at 4℃ for 5min, and then transfer the supernatant into a new tube.

5. Resuspend pellet of 10mL bacterial culture (or 100mL bacterial culture for very low expression level) in 1mL lysis buffer.

6. Prepare lysis buffer containing urea 6~8mol/L or Guanidine-HCl 6mol/L (try 8mol/L of urea first, and if protein is soluble titer down in the next experiments till minimal urea is required for protein solubilization).

7. Sonicate on ice 20s 3 times (depends on the sonicator). Spin 15min max speed at 4°C.

8. Transfer supernatant into clean tube: crude extract (keep 40μL for SDS-PAGE).

9. Equilibrate 50μL Ni-NTA agarose beads with equilibration buffer: place 50μL beads (100μL suspension) of Ni-NTA agarose beads in 1.5mL plastic tube. Wash 2 times respectively with 1.5mL H$_2$O and 1.5mL equilibration buffer (washing: mix, spin 3,500r/min×3min, discharge supernatant).

10. Add the crude extract to the beads and incubate 4℃ for 1h (swirl).

11. Spin 3,500r/min×3min. Discharge unbound material (keep 40μL for SDS-PAGE).

12. Wash 3 times with 1mL wash buffer (appropriate urea concentration). Washing: mix, spin 3,500r/min×3min, and discharge supernatant (keep 40 μL for SDS-PAGE).

13. Wash 2 times with 1mL wash buffer + 10mmol/L imidazole (keep 40 μl for SDS-PAGE).

14. Elute 2 times with 100μL elution buffer + 100mmol/L imidazole (keep 40μL for SDS-PAGE).

15. Final elution: 2 times with 100μL 250mmol/L imidazole (keep 40μL for SDS-PAGE).

16. Run on SDS-PAGE gel 5μL of crude extract and unbound material, and 13μL of the wash and elution fractions.

Notes

1. Do not expose Ni matrices to reducing agents as DTT (you can use β-mercaptoethanol up to 20mmol/L); chelating agents as EDTA and EGTA; NH$_4^+$ buffers and amino acids as Arg, Glu, Gly or His.

2. Alternative protocol if target protein is not pure enough: Perform parallel purification procedures where you include 10, 20, 30, 40 or 50mmol/L imidazole in the lysis, binding and washing buffer. Elute directly 3 times

with 100μL elution buffer + 250mmol/L imidazol. Check eluted proteins on SDS-PAGE. Expect lower yields but higher purification by increasing the imidazol concentration.

Part V Cleavage of His-tagged Proteins with Thrombin Cleavage Sites

Principle

The fusion gene expressive system has been widely used to produce exogenous proteins in prokaryotic cells. This is completely due to that these systems can generate large amounts of soluble fusion protein. With the universal application of the fusion protein expression, it is particularly important that N-terminal of fusion protein is removed from C-terminal of the target protein. Currently, there are two methods for lysis of the fusion protein in specific site. Chemical method is cheaper and more effective, even can lyse proteins under the denaturing conditions, but it has poor specificity. Enzymatic hydrolysis is more moderate, and highly specific.

Thrombin recognizes the consensus sequence Leu-Val-Pro-Arg-Gly-Ser, cleaving the peptide bond between Arg and Gly. This is utilized in many vector systems which encode such a protease cleavage site allowing removal of an upstream domain. Predominantly the domain to be cleaved is a purification tag such as a "His-tag".

Reagents and material

1. Low-medium salt buffer (20mmol/L Tris-Cl, 100mmol/L NaCl at pH8, no protease inhibitors).
2. 1% thrombin (purchase from Calbiochem).
3. Benzamidine Sepharose column (purchase from Pharmacia).
4. PMSF (1mmol/L, using a freshly prepared 100mmol/L ethanolic stock solution).
5. Benzamidine (0.1~1mg/mL from a 50mg/mL aqueous solution).
6. The reagents for SDS-PAGE.

Procedure

1. After the Ni-column purification it is advisable to exchange (e.g., dialyse) the protein into a low-medium salt buffer. High imidazole may inhibit the cleavage.

2. Add 1% thrombin preferably at room temperature and shake gently for 1h (use the highest grade human thrombin).

3. (Optional) Monitor reaction the first time by taking aliquots (10μL) at various time intervals and, via SDS-PAGE look for a gel shift—the protein is 2kU lower in molecular weight when cleaved so is easily detectable (provided your protein is in the 0~40kU range). Keep some uncleaved protein as a control and spot a mixture of uncleaved and putatively cleaved in one lane. You may need to allow the gel front to run into the gel buffer solution to get sufficient resolution.

4. Separate the cleaved protein from any uncleaved protein and the His-containing peptide by running the whole mixture through another Ni-NTA column under the same conditions as the first time. The cleaved protein should just flow through and uncleaved protein (which you may wish to recover) will bind to the column and can be eluted with imidazole.

5. Run the collected fraction through a pre-equilibrated (in buffer of choice and with 20 column vol. of equilibration) Benzamidine Sepharose column (1mL of slurry). This should bind residual thrombin preventing secondary activity from this source and some others. Collect the flow through.

6. Add 1mmol/L PMSF and 0.1~1mg/mL benzamidine and any other protease inhibitors you deem necessary.

7. Identified by SDS-PAGE.

Notes

A problem with many commercial thrombin sources is secondary protease activity. If this were due to thrombin alone then this would not, in general, be an insurmountable problem, since this can relatively easily be dealt with (with benzamidine and via a benzamidine sepharose column). However human thrombin is one of the most active site-specific proteases. Our experience suggests that such secondary activity arises from other proteases present in the purchased thrombin. Some groups generally further purify purchased thrombin with a Mono-Q ion exchange chromatography step which is an advisable and simple further step.

Section III Integrative Experiments

Experiment title

Date
Observations and results

Discussion

1. Expression vectors require not only _____ but also _____ of the cloned gene, thus require more components than simpler cloning vectors.

2. High levels of expression of recombinant proteins in a bacterial system can lead to the formation of _____.

3. What is the difference between prokaryotic expression system and eukaryotic expression system?

4. What is the advantage if the foreign gene is expressed as fusion protein?

Teacher's remarks
Signature
Date

Chapter 12 Innovation Experiment Design and Application

The basic aim of biochemistry and molecular biology is to analyze the relationship between the structure and function of biological macromolecules. Experimental techniques in biochemistry and molecular biology can be divided into three levels according to research objectives and characteristics. The first level is the qualitative and quantitative identification and analysis of biological macromolecules, the second level is the artificial synthesis or reconstruction of biological molecules, and the third level is the functional study of biomolecules.

At present, it has formed a complete set of research methods in the field of biochemistry and molecular biology. With grasping of these experimental techniques and preforming the sophisticated design and use of various techniques, we have the ability to isolate, prepare, analyze and identify the unknown biological molecules; to study the physicochemical properties, functional active sites, molecular structures and intermolecular interactions of biological molecules; to innovate biomolecules by genetic engineering, including design and construction of recombinant DNA, cloning and efficient expression of gene products (RNA, peptides and proteins). The targeted mutation of gene can produce the proteins with special functions, so as to carry out targeted metabolic and functional innovation at the cell or individual level. Therefore, innovative experimental design will lay a foundation for the study of molecular mechanism, diagnosis and therapy of diseases in the field of medicine.

This chapter selects several common research examples to show the basic procedures of designing innovative experiments. It includes selection of a subject, design of experimental technical route and selection of appropriate techniques and methods; implementation of experiments, observation and analysis of experimental progress, summary of experimental problems and results. It helps us to gradually transition from the mastery of basic experiments to the understanding and design of comprehensive experiments, so as to facilitate the development of innovative research in the future.

Experiment 19　Separation and Identification of Alkaline Phosphatase

A comprehensive experiment on the extraction of alkaline phosphatase (AKP or ALP) from rabbit liver, provides a design experimental scheme for the separation and purification of enzyme proteins. The basic techniques and methods include quantification by spectrophotometry, separation using precipitation, centrifugation, chromatography, and identification of target protein purity and determination of specific enzyme activity.

Design and Principle

AKP is an enzyme that has low substrate specificity and can hydrolyze various phosphate monoester compounds in an alkaline environment. It has an optimum pH range of 8.6~10 and requires Mg^{2+} or Mn^{2+} as activators. The characteristic activity of AKP is to transfer the phosphate group of substrate to the hydroxyl group of acceptor (Such as H_2O). AKP is mainly found on the cell membranes of small intestinal mucosa, kidney, bone, liver and placenta. Serum AKP is mainly from the liver and also from bone. The separation and purification method of AKP is similar to general protein which purified by salting out, electrophoresis, chromatography and organic solvent precipitation.

In this experiment, rabbit liver is selected for AKP extraction by stepwise precipitation of organic solvent. AKP is extracted with 30% ethanol and 33% acetone, and the precipitate is discarded after centrifugation to remove impurities and impurity proteins. Then the supernatant is extracted with 60% ethanol and 50% acetone, AKP is insoluble in 60% ethanol and 50% acetone. After centrifugation, the precipitate is reserved to obtain a pure AKP.

The specific activity of an enzyme refers to the unit of enzyme activity per milligram of the enzyme. As the enzyme is gradually purified, its specific activity is gradually increased. Therefore, the specific activity can be used to identify the purity degree of the enzyme. Determining the enzyme specific activity by measuring the milligrams and the activity unit of per milliliter of the enzyme sample. AKP activity was determined by the phenylphosphonium phosphate method, using sodium phenyl phosphate as a substrate, which was hydrolyzed by AKP to produce free phenol and phosphate. The phenol can react with 4-aminoantipyrine in an alkaline solution, and be oxidised by potassium ferricyanide to produce a red indole derivative. The activity of the enzyme can be measured the production quantitation according to depth of the red color. The AKP content can be determined by BCA method. Finally, the specific activity is calculated based on the measured content and the number of activity units of enzyme, and the degree of purification of the enzyme is also identified.

Material and reagents preparation

1. 0.5mol/L magnesium acetate ($MgAc_2$) solution: magnesium acetate 10.73g is dissolved in ddH_2O, diluted to 100mL.

2. 0.1mol/L sodium acetate (NaAc) solution: NaAc 8.2g are dissolved in ddH_2O and diluted to 1L.

3. 0.01mol/L $MgAc_2$-NaAc mixed solution: 0.5mol/L $MgAc_2$ solution 20mL, 0.1mol/L NaAc solution 100mL, are mixed and diluted to 1L with ddH_2O.

4. Tris-$MgAc_2$ buffer (pH8.8): 13.1g of Tris are dissolved in ddH_2O, and diluted to 1L to form a 0.1mol/L Tris solution. Then, 0.1mol/L Tris solution 100mL, 0.1mol/L $MgAc_2$ solution 100mL, and mixed with 800mL of ddH_2O, adjust the pH to 8.8 with 1% acetic acid, and diluted to 1L with ddH_2O.

5. 0.1mol/L carbonate buffer (pH10): Na_2CO_3 6.36g and $NaHCO_3$ 3.36g are dissolved in ddH_2O and diluted to 1L.

6. Substrate solution: phenyl phosphate disodium dehydrate 6g, 4-aminoantipyrine 3g, are respectively

dissolved in ddH$_2$O which was boiled and cooled down, and the two solutions are mixed and diluted to 1L. Store in a brown bottle in refrigerator for one week. This solution can be mixed in an equal amount with 0.1mol/L pH10 carbonate buffer at the time of use.

7. 5g/L potassium ferricyanide solution: potassium ferricyanide 5g, boric acid 15g, are dissolved in 400mL ddH$_2$O respectively, then they are mixed together and diluted to 1L.

8. Phenol standard solution: 1.5g of recrystallized phenol (AR grade) are dissolved in 0.1mol/L HCl up to 1L to obtain stock solution. 25mL of the above phenol solution are taken in 250mL iodine flask, added 50mL 0.1mol/L NaOH and heated to 65℃, then added 0.1mol/L iodine solution 25mL, cover the iodometric bottle stopper, and placed it for 30min. 5mL of concentrated hydrochloric acid are added in, titrated with 0.1mol/L standard sodium thiosulfate with 0.1% starch solution as an indicator. So 0.1mol/L iodine solution of per milliliter (12.7mg iodine) equivalent to the mg number of phenol is: $12.7 \times 94/3 \times 254 = 1.567$. Assuming that the remaining iodine after the action of 0.1mol/L iodine solution 25mL and 25mL phenol solution is determined to be X mL with Na$_2$S$_2$O$_3$, the amount of phenol contained in the 25mL phenol solution is $(25-X) \times 1.567$ mg, which is calculated phenol stock solution concentration. Finally, the stock solution was diluted to 0.1mg/mL application solution.

9. Acetone.

10. Protein standard (0.2mg/mL): bovine serum albumin (BSA) 20mg are dissolved in 0.9% NaCl solution and diluted to 100mL.

11. BCA reagent: purchase from commercial company.

Procedure

1. Isolation and purification.

(1) Homogenization: Weigh 2g of fresh rabbit liver and put it into a glass homogenizer, add 6mL of 0.01 mol/L MgAc$_2$-NaAc mixed solution, homogenate in a homogenizer for 3~4min, pour homogenate into the graduated centrifuge tube and record the volume. Take 0.1mL homogenate in another tube and add 4.9mL of pH8.8 Tris-MgAc$_2$ buffer to dilute it, which is marked as the A solution to be tested for specific activity.

(2) Extraction: Add 2mL of n-butanol to the homogenate, stir well with a glass rod for 2min, place at room temperature for 30min, filter with gauze and place the filtrate in a graduated centrifuge tube.

(3) Acetone precipitation: Add an equal volume of cold acetone to the filtrate, mix and centrifuge at 3,000r/min for 5min, discard the supernatant, add 4mL of 0.5mol/L MgAc$_2$ solution to the precipitate, and stir well with a glass rod, record the volume of the suspension. Take 0.1mL suspension in another tube and mix with pH8.8 Tris-MgAc$_2$ buffer 4.9mL, which is marked as B solution to be tested for specific activity.

(4) Step separation: add 95% cold ethanol to the suspended suspension to make the final volume fraction of ethanol to 30%. After mixing, immediately centrifuge at 2,000r/min for 5min. Pour the supernatant into another centrifuge tube. Discard the precipitate. The supernatant was mixed with cold 95% ethanol to make the final volume fraction of ethanol to 60%. After mixing, centrifuge at 3,000r/min for 5min. The supernatant was discarded. 4mL of 0.01mol/L MgAc$_2$ was added to the precipitate, stirred well to completely suspend.

(5) Repeat the above procedure. After the supernatant was discarded, the precipitate was fully suspended with 0.5mol/L MgAc$_2$ solution 3mL, and record the volume. Take 0.1mL and mix with pH8.8 Tris-MgAc$_2$ buffer 0.9mL, marked as the C solution to be tested for specific activity.

(6) Cold acetone was added to the remaining suspension to make the final volume fraction of propionate to 33%. After mixing, centrifuge at 2,000r/min for 5min, and discard the precipitate. Add cold acetone to the supernatant to make the final volume fraction of acetone to 50%, centrifuge at 3,000r/min for 15min, discard the supernatant. Add 5mL of Tris-MgAc$_2$ buffer to the precipitate, after mixing well, centrifugation for 5min at

3,000r/min, the supernatant is the purified enzyme solution and marked as the D solution to be tested for specific activity.

2. AKP activity determination by phenyldiphenyl phosphate method.

(1) Take six test tube to label and operate according to the following table:

Reagents (mL)	Blank	Standard	A	B	C	D
Diluted enzyme solution in each stage	—	—	0.1	0.1	0.1	0.1
Enzyme standard solution (0.1mg/mL)	—	0.1	—	—	—	—
Tris-MgAc$_2$ buffer (pH8.8)	0.1	—	—	—	—	—
Substrate solution (preheat to 37℃)	3.0	3.0	3.0	3.0	3.0	3.0

(2) After add the substrate solution, mix immediately, and hold in water bath at 37℃ for 15min (accurately control time). Then add 0.5mL of 0.5% potassium ferricyanide solution to each tube, mix immediately to stop the enzymatic reaction. Measure the A_{510nm} after standing for 15min.

(3) Enzyme activity unit calculation: one active unit for AKP activity is defined as every 1mg phenol produced at 37℃ for 15min. Therefore, the enzyme activity unit per milliliter of enzyme solution is as follows:

$$\text{AKP activity unit} = \frac{A_{sample}}{A_{standard}} \cdot \frac{C_{phenol\ standard}}{0.1} \cdot \text{dilution factor}$$

3. AKP protein content determination.

(1) Take 6 test tube labels and operate according to the following table:

Reagents (mL)	Blank	Standard	A	B	C	D
Diluted enzyme solution in each stage	—	—	0.1	0.1	0.1	0.1
Tris-MgAc$_2$ buffer (pH8.8)	0.1	—	—	—	—	—
Protein standard solution (0.2mg/mL)	—	0.1	—	—	—	—
BCA reagent	1.5	1.5	1.5	1.5	1.5	1.5

After mixing, the tubes was heated at water bath at 37℃ for 30min, and the protein concentration was calculated by A_{562}nm within 1h.

(2) Protein concentration calculation:

$$C_{AKP\ sample} = (A_{sample}/A_{standard}) \cdot C_{standard} \cdot \text{dilution factor or } \frac{A_{sample}}{A_{standard}} \cdot C_{standard} \times \text{dilution factor}$$

4. Specific activity calculation:

AKP specific activity = AKP activity unit /mL sample or AKP activity unit /mg protein sample

5. Calculate the purification factor and yield of AKP at each stage.

Separation phase	Volume	Protein (mg/mL)	Enzyme activity(U/mg)	Specific activity(U/mL)	Purification factor	Yield
A						
B						
C						
D						

Notes

1. The final concentration of the organic solvent in the extraction process should be calculated correctly and added accurately. Organic solvents must be precooling.

2. It should be clear whether the supernatant or precipitation is retained after each centrifugation.

3. When measuring protein concentration in each step of separation and purification, attention should be paid to dilution ratio.

Experiment 20 Lactate Dehydrogenase (LDH) Analysis

Using lactate dehydrogenase (LDH) as an example, a comprehensive experiment is to illustrate how to prepare an enzyme by isolating it from fresh tissues or cloning it and how to determine activity, purity, molecular weight, subunit composition, and the enzyme kinetics of an enzyme.

This experiment will be carried out in accordance with the relevant principles and protocols mentioned above. Design a workflow and prepare a more detailed protocol and reagents. The researcher must record the experiment process, observation, data, calculation and analysis in the experiment report.

Part I Purification of Lactate Dehydrogenase (LDH) from Chicken

Design and Principle

LDH, an enzyme converting pyruvate to lactate in glycolysis, is found in nearly all living cells. LDH is a tetramer containing two types of subunits, M subunit and H subunit, which can form five tetramers (isoenzymes): 4H, 3H1M, 4M, 2H2M, 1H3M. LDH-M is encoded by the LDHA gene which is expressed more in muscle and liver. LDH-H is rich in heart and encoded by the LDHB gene.

In order to study the properties of LDH, LDH inside the cells should be separated and purified from the tissue. The most efficient LDH purification protocol is set up based on the differences of properties between LDH and other molecules. Major steps of LDH purification shown here include tissue homogenization and ammonium sulfate fractionation, desalting by gel filtration and affinity chromatography (Fig. 12-1). During the process of enzyme purification, protease inhibitors and low temperatures (4°C or on ice) are commonly used to keep proteolysis to a minimum.

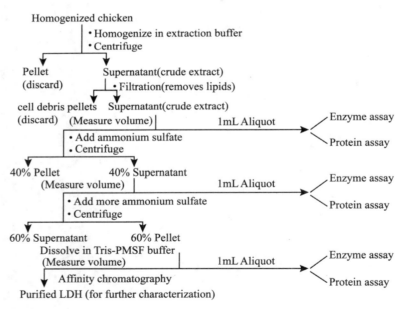

Figure 12-1 Flow chart for isolation of LDH

* **Extraction Buffer**: 10mmol/L Tris-HCl (pH7.4), 1mmol/L 2-Mercaptoethanol, 1mmol/L Phenylmethylsulfonyl fluoride (PMSF) and 1mmol/L Ethylenediamine tetraacetic acid (EDTA); * **Tris-PMSF buffer** 10mmol/L Tris-HCl (pH8.6), 0.5mmol/L 2-Mercaptoethanol and 1mmol/L PMSF.

Part II LDH Activity assays

The LDH activity will be determined on the following samples: crude homogenate, Desalted ammonium sulfate fraction, and peak fraction(s) from the Cibacron Blue column.

LDH is an oxidoreductase which catalyzes the interconversion of lactate and pyruvate, the equation is:

$$\text{Lactate} + \text{NAD}^+ \xleftrightarrow{\text{LDH}} \text{Pyruvate} + \text{NADH} + \text{H}^+$$

In order to the measure the concentration of a specific enzyme present in a sample, you can either measure the reaction rate catalyzed by the specific enzyme directly or measure the amount of product catalytically produced by the enzyme per unit time under a specified set of conditions. Activity of serum LDH was measured by determining the NADH production or pyruvate production using Spectrophotometry.

The amount of LDH in a sample can be determined by measuring the initial reaction rate at which NADH is consumed when substrate concentration is much higher than enzyme concentration. When adding a sample into the UV cuvette containing substrates, the change in A_{340nm} should be proportional to the change in NADH concentration due to the LDH activity present. The spectrophotometer will calculate rates in units of change in A_{340nm} per unit time ($\Delta A/\min$). These values will then be converted to $\mu mol/\min$ using the ε_{340} for NADH = 6220 $\mu mol \cdot (\mu L \cdot cm)^{-1}$.

One unit of LDH activity can also be defined as the amount of enzyme needed to produce 1.0 μmol of pyruvate per 100mL sample after it is incubated with the substrate at 37℃, pH10.0 for 15min. the reagent, 2,4-dinitrophenylhydrazine, is added for preventing the reaction and marron compounds are formed which response to pyruvate. Read A_{340nm} using a spectrophotometer. The intensity of the color formed is directly proportional to the enzyme activity.

Part III LDH Protein Assay

During LDH purification, the LDH activity and total protein concentration in each step will be determined to evaluate the effectiveness of the procedure. Different methods for measuring the amount of protein are appropriate here, including the Bradford assay.

A purification table can be set up to measure the removal of contaminants during the purification, and to assess the efficiency of the purification process.

LDH Purification Table

Step	Total enzyme activity	Total protein concentration	Specific activity	Purification fold	Yield

Specific activity: (enzyme activity / protein concentration);
Purification-fold: (specific activity at a given step) / (specific activity of starting sample);
Yield: (Total activity at a given step) / (Total activity of starting sample) ×100.

Part IV LDH Analysis Using SDS-PAGE and Western Blotting

Once LDH purification procedure has been completed, the resulting protein must be characterized for both purity and physical properties including MW and subunit composition. SDS-PAGE allows to assessment of purity of the LDH sample, and measurement of the size of the protein existed. The MW of protein, which can either refer to that of the intact multisubunit protein, or to that of each of its individual subunits. The most common method for determining the subunit MW of a protein is by PAGE under denaturing conditions. To obtain the MW of the intact multisubunit protein, PAGE should be done under native conditions.

You might need to run all purified protein samples that contain LDH activity, including crude extract, desalted ammonium sulfate pellet and Cibacron Blue NADH elution. After running SDS-PAGE, one half will be stained by Coomassie brilliant blue and the other half will be for identification of LDH using Western blotting.

Part V Catalytic properties of the LDH

Enzyme kinetics of the purified LDH can be further characterized, including determining K_m and V_{max} values using pyruvate as a substrate and studying the effects of an inhibitor on LDH activity. In order to determine the K_m and V_{max} values of the LDH preparations, enzyme assays should be carried out in the presence of increasing concentrations of substrate. The initial rate (V_i) determined from each assay is then plotted vs. the corresponding initial substrate concentration to obtain the Lineweaver-Burk plot, from which the K_m and V_{max} can be evaluated.

The study of enzyme kinetics in the presence of specific inhibitors can be very informative. Oxamate, a close structural analog of pyruvate, competes directly and effectively with pyruvate for binding to the LDH active site. To determine the type of inhibitor, the K_m and V_{max} must be determined in the absence of inhibitor, and in the presence of more than one concentration of inhibitor. The results can be summarized with headings as shown in the following table. The Lineweaver-Burk plot can then be constructed based on the results and the inhibition type can be indicated from the plot.

Report: Catalytic properties of LDH

Tube			No inhibitor		0.04mM inhibitor		0.2mM inhibitor	
Tube#	[S]	1/[S]	V_i	$1/V_i$	V_i	$1/V_i$	V_i	$1/V_i$

Part VI Cloning of Human LDH

Bacterial expression system generally allows a large number of exogenous proteins to be expressed, and can choose the form of the protein expression, and allows researchers to introduce mutations in protein sequences to examine the role of individual amino acids in protein function. Human LDH A subunits can be cloned and expressed in *E. coli* to obtain LDH proteins which is exclusively composed of the LDH A isoenzymes.

The general outline of the procedure is shown below.

1. PCR and plasmid preparation.

The primers were designed to clone the entire coding region (with specific restriction sites at each end) of the human LDH A subunits:

5′primer: TCCAccATGGCAACTCTAAAGGATCAG

3′primer: CAGAAagCTTTAAAATTGCAGCTCC

The lowercase bases in the primers are mismatches to generate specific restriction sites (the underlined sequences). The 5′ primer contains an *Nco* I site and the 3′ primer contains a *Hind* III site.

2. Plasmid preparation: the map of the expression plasmid pTrc99A, which contains an *Nco* I site and a *Hind* III site, is shown in Fig. 12-2.

3. Restriction endonucleases (REs) digest on plasmids and PCR products, then separate the target DNA

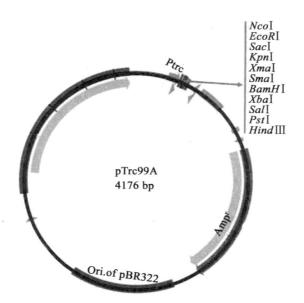

Fig. 12-2 The Vector map of pTrc99A

fragments of plasmids and PCR products by agarose gel electrophoresis, respectively, and purify DNA fragments using DNA Gel Extraction Kit.

4. Ligation and transformation.
5. Selection and screening of colonies for protein expression.
6. Enzymatic activity measurement of expressed LDH.

… Section III Integrative Experiments

Experiment 21　Protein Markers Exploration by Two-dimensional Electrophoresis

Using the basic principle and technique of two-dimensional electrophoresis (2-DE), all proteins in the cell can be displayed at the proteome level, which is conducive to exploring proteins that are different from normal and disease states, and laying a foundation for exploring protein markers of disease.

Design and Principle

2-DE is a method for separating proteins by combining Isoelectric Focusing (IEF) electrophoresis and SDS-PAGE. It has a very high resolution at 1,000~3,000 points, called "protein explosion", which is the key basic technique for proteomics.

2-DE with binding IEF and SDS-PAGE is established by O'Farrall, et al. in 1975, the method is based on differences in pI and MW between different components. IEF electrophoresis is the first direction and SDS-PAGE is the second direction. In the IEF system, a high concentration of urea and an appropriate amount of non-ionic detergent NP-40 should be added in. In addition to these two substances, dithiothreitol (DTT) should be used to denature protein and make peptide chain stretching. After the end of IEF electrophoresis, the IPG strip was shaken and equilibrated in the sample treatment solution (containing SDS, DTT, iodoacetamide) for SDS-PAGE, and embedded on the gel plate end of SDS-PAGE. Then, 2-DE can be performed. 2-DE is extremely delicate for the separation of proteins (including ribosomal proteins, histones, etc.) Therefore it is particularly suitable for isolating complex protein components in cells.

Materials and Reagents

1. IEF system.

(1) IPG dry strip (pH3~10, 18cm).

(2) Urea (ultra-pure grade).

(3) Ampholine ampholyte (pH3~10).

(4) DTT (dithiothreitol).

(5) Cell lysis Solution: 7mol/L Urea, 2mol/L thiourea, 4% CHAPS, 2% pharrmalyte pH3~10 (add before use), 1% DTT (add before use), protease inhibitor (add before use), 40mmol/L Tris-base.

(6) Hydration solution: 8mol/L urea, 2% CHAPS, 0.5% IPG buffer, 1% DTT (add before use), 0.002% bromophenol blue.

(7) Equilibration buffer: 50mmol/L Tris-Cl, 6mol/L Urea, 2% SDS, 1% DTT (add at first equilibration), 4% bromoacetamide (add at second equilibrium addition), 30% glycerol, 0.002% bromophenol blue.

2. SDS-PAGE system, see Experiment 5.

3. Sample preparation system: 10% trichloroacetic acid (dissolve in acetone, containing 0.1% DTT).

4. Silver dyeing system.

(1) Fixing solution: 500mL ethanol, 100mL glacial acetic acid, 400mL ddH$_2$O.

(2) Soaking solution: 75mL ethanol, 17g sodium acetate, 1.25mL 25% glutaraldehyde (fresh preparation), 0.5g sodium thiosulfate pentahydrate.

(3) Silver staining solution: 0.25g silver nitrate and 50μL formaldehyde (freshly prepared) were added to 250mL with ddH$_2$O.

(4) Coloring solution: 6.25g sodium carbonate and 25μL formaldehyde (freshly prepared) were added to 250mL with ddH$_2$O.

(5) Stop solution: 3.65g dihydrate EDTA-Na$_2$ was added to 250mL with ddH$_2$O.

(6) Preservation solution: 25mL 87% glycerol was added to 250mL with ddH$_2$O.

Procedure

1. Sample processing.

(1) After the rats are sacrificed, the liver is quickly taken out and washed with precooled ddH$_2$O.

(2) The liver is homogenized by adding five-fold volumes of precooled 10% trichloroacetic acid (TCA) in an ice water bath.

(3) Place at −20℃ for 1h and centrifuge at 40,000g for 30min.

(4) The protein precipitate is rinsed twice with an acetone solution containing 0.1% DTT to remove residual TCA.

(5) The appropriate amount of urea solution is added to fully dissolve the protein precipitate.

(6) Centrifuge at 40,000g for 30min, take the supernatant, quantify by Bradford method, adjust the protein concentration to about 4mg/mL, and take 150μg protein sample for 2-DE.

2. First-direction IEF electrophoresis.

(1) Add a proper amount of hydration solution (18cm strip, total volume of 350μL) to the protein sample and place it in a standard strip groove.

(2) Strip the protective film of the IPG strip from the acid end, with the rubber surface facing down, and slowly put the strip into the strip groove (acid end toward the tip), and the hydration solution wets the entire strip. Ensure the ends of the strip in contact with the electrodes at both ends of the tank, taking care to avoid air bubbles.

(3) Cover the IPG strip with a proper amount of oil and cover it.

(4) Place the strip groove on the IPGphor instrument for electrophoresis, paying attention to the direction of the electrode and full contact with the platform.

(5) Set the IPGphor instrument operating parameters as shown in the table below.

Parameter	Set value
temperature	20℃
Maximum current	0.05mA / IPG strip
Sample volume	350μL (18cm strip)
Voltage	
30V	10~12h (Hydration)
200V	1h
500~800V	0.5h
800V	3h

(6) IEF can be performed automatically after IPG strips are hydrated.

3. IPG strip balance.

Turn off the power, end the electrophoresis, remove the strip, add 10mL of equilibration solution (contain DTT, the denatured non-alkylated protein is in a reduced state), shake for 15min, remove the equilibration solution, and add 10mL of equilibration solution containing iodoacetamide (to close the sulfhydryl group) and shake for 15min to remove the equilibration solution.

4. Second direction SDS-PAGE: prepare a gel solution according to Experiment 5.

(1) Aspirate the covered ddH$_2$O with a filter paper, and add 1% hot agarose to the top of the glass plate.

Meanwhile, the IPG strip was carefully placed on top of the slab gel and solidified with 1% hot agarose gel (prepared with an equimolar buffer without bromophenol blue).

(2) Add the electrode buffer solution to the upper and lower tanks, and connect the electrophoresis tank to the electrophoresis apparatus with a wire, the upper tank is connected to the negative electrode, and the lower tank is connected to the positive electrode. After constant pressure 80V electrophoresis until the sample entered the polyacrylamide gel, a constant pressure of 140V was electrophoresed until the indicator dye reached the bottom of the gel.

(3) Remove the electrophoretic glass plate, carefully pry open the glass plate, separate the two glass plates (to prevent breaking the gel and the glass plate), draw the gel and place it in the enamel pan and then silver staining.

5. Silver dye.

(1) Fixation: fix in the fixative for at least 30min.

(2) Soaking: soak in the soaking solution for 30min.

(3) Rinsing: rinse with ddH_2O for 3 times, 5min/time.

(4) Silver stain: stain in silver staining solution for 20min.

(5) Color development: color development in the coloring solution for 2~10min, visual protein band color is dark brown.

(6) Termination: soak in the stop solution for 10min.

(7) Rinsing: rinse with ddH_2O for 3 times, 5min/time.

(8) Preservation: soak in the preservation solution for 30min, remove and dry.

Precautions

(1) The choice of ampholyte carrier is mainly based on the approximate pI range of the protein sample.

(2) The sample should be desalted, otherwise the zone is twisted; to dissolve completely, the impurity precipitate can be removed by centrifugation, and the undissolved particles are easy to cause tailing; the amount of sample depends on the type of protein in the sample and the sensitivity of the detection method. It is optimal that the loading volume is 20~40μL and the concentration is 1.5~3μg/μL. The sample is subjected to pre-electrophoresis.

(3) The electrode buffer should be selected according to the pH range of the ampholyte.

(4) Micro-injector, syringe and pipette should be drained and flushed immediately to prevent solidification blockage.

(5) Many experimental equipment should be carefully clean and put back in place.

Reference

[1] Audagnotto M, Dal Peraro M. Protein post-translational modifications: *In silico* prediction tools and molecular modeling[J]. *Comput Struct Biotechnol J*, 2017, 15: 307-319.

[2] Chu C, Chang HY. Understanding RNA-Chromatin interactions using chromatin isolation by RNA purification (ChIRP)[J]. *Methods Mol Biol*, 2016, 1480: 115-123.

[3] Chuh KN, Batt AR, Pratt MR. Chemical methods for encoding and decoding of posttranslational modifications [J]. *Cell Chem Biol*, 2016, 23(1): 86-107.

[4] Ebert MS, Neilson JR, Sharp PA. MicroRNA sponges: competitive inhibitors of small RNAs in mammalian cells[J]. *Nat Methods*, 2007, 4: 721-726.

[5] Han X, Aslanian A, Yates JR III. Mass spectrometry for proteomics[J]. *Curr Opin Chem Biol*, 2008, 12 (5): 483-490.

[6] Huang J, Wang F, Ye M, Zou H. Enrichment and separation techniques for large-scale proteomics analysis of the protein post-translational modifications[J]. *J Chromatogr A*, 2014, 1372C: 1-17.

[7] Hubert Rehm. Protein Biochemistry and Proteomics[M]. Oxford: Elsevier Inc., 2006.

[8] Johnson D S, Mortazavi A, Myers R M, Wold B. Genome-wide mapping of in vivo protein-DNA interactions [J]. *Science*, 2007, 316(5830): 1497-1502.

[9] Jocelyn E Krebs, Elliott S. Goldstein, Stephen T. Kilpatrick. Lewin's Genes XI[M]. New York: Jones & Bartlett Publishers, 2014.

[10] Julie D Thompson, Christine Schaeffer-Reiss, Marius Ueffing. Functional proteomics: methods and protocols [M]. New York: Humana Press, 2008.

[11] Katia M. Arndt, Kristian M. Muller. Protein engineering protocols[M]. New York: Humana Press, 2011.

[12] Michael R. Green. Molecular cloning: a laboratory manual (4th Edition)[M]. New York: Cold Spring Harbor Laboratory Press, 2012.

[13] Nawin Mishra. Introduction to Proteomics. Principles and Applications[M]. Hoboken: John Wiley & Sons, 2010.

[14] Rao VS, Srinivas K, Sujini GN, Kumar GN. Protein-protein interaction detection: methods and analysis[J]. *Int J Proteomics*, 2014: 147648.

[15] Robert F. Weaver. Molecular Biology (5th Edition)[M]. Hoboken: McGraw Hill, 2012.

[16] Robert K Murry, David A Bender, Kathleen M Botham, Peter J Kennelly, et al. Harper's illustrated Biochemistry (28th Edition)[M]. Hoboken: McGraw Hill, 2009.

[17] Richard R Burgess, Murray P Deutscher. Guide to Protein Purification (2nd Edition)[M]. New York: Academic Press, 2009.

[18] Yu Hong, Huang XinXiang. Experimental manual in medical biochemistry (1st Edition)[M]. Wuhan: Wuhan University Press, 2008.

[19] Umeyama T, Ito T. DMS-seq for in vivo genome-wide mapping of protein-DNA interactions and nucleosome centers[J]. *Curr Protoc Mol Biol*, 2018, 123(1): e60.

[20] Van Rooij E, Sutherland L B, Qi X, Richardson J A, Hill J, Olson E N. Control of stress-dependent cardiac growth and gene expression by a microRNA[J]. *Science*, 2007, 316 (5824): 515-579.